HOUSE PLANTS

Reaktion's Botanical series is the first of its kind, integrating horticultural and botanical writing with a broader account of the cultural and social impact of trees, plants and flowers.

HOUSE PLANTS

Mike Maunder

REAKTION BOOKS

Published by
REAKTION BOOKS LTD
Unit 32, Waterside
44–48 Wharf Road
London N1 7UX, UK
www.reaktionbooks.co.uk

First published 2022
Copyright © Mike Maunder 2022

Printed and bound in India by Replika Press Pvt. Ltd

A catalogue record for this book is available from the British Library

ISBN 978 1 78914 543 4

Contents

𖣯

The relationship between people and house plants is both ancient and intimate, as seen in this 15th-century illustration of a bedroom with potted plant, from *Der Trojanische Krieg* (1455).

Introduction
Plants of the Indoor Biome

🪴

'The chief end of labor should be human happiness, and so the
effort that is put forth in the cultural art of houseplants not
only brings happiness to the heart of the grower, but also to
the passer-by who with a hungry soul admires this plant.'

HUGH FINDLAY, 1916[1]

This is an exploration of an apparently mundane group of
plants, the house plant. Like the relationship we have with
edible and medicinal plants, our bond with house plants is
an intimate botanical relationship; after all, we choose to bring them
into our homes. Whether a thriving and diverse collection loved by
its owner, or a chlorotic embarrassment, house plants tell a complex
story about how we live, why we need nature and how we take wild
things and domesticate them.

Much has been written about plant blindness, people's seeming
ability to overlook and undervalue plants, to the extent that vital
subjects such as crop diversity and plant conservation are profoundly
under-resourced. However, the fact that house plants exist, are bought
by the million and bring joy to many suggests that a large proportion
of our species is not plant blind, and that is a very good thing.

Our love for house plants opens a door to exploring important
stories about plant domestication and perhaps ultimately about the
mutualism between plants and people. Some house plants exist un-
modified, almost as they were when collected from the wild, for

7

Perhaps the ultimate house plant, the African violet was changed profoundly by a century of breeding. Illustration from *Curtis's Botanical Magazine*, vol. CXXI (1895).

instance the spectacular Swiss cheese plant, *Monstera deliciosa*. Others, such as those ultimate house plants the African violet, *Saintpaulia*, and the moth orchid, *Phalaenopsis*, have been profoundly altered by breeders over many decades, and tell a story of domestication, in which art and science have worked together.

We are becoming an urban species, and more of our existence than ever is spent indoors insulated from nature and living in an increasingly impersonal world. Fewer people have gardens, and those who rent a property tend not to invest their time in cultivating a garden. At the same time, we are told we possess a deep relationship with nature that is essential for our well-being. The naturalist and writer E. O. Wilson defined this as 'biophilia', the 'connections that human beings subconsciously seek with the rest of life'.[2] Sadly, much of our history has shown that our relationship with life has been manifest through the extermination of species and the conversion of wild lands into domesticated landscapes. However, a recent and

An investment in science changed the moth orchid, *Phalaenopsis*, from expensive luxury to supermarket commodity. Illustration by John Nugent Fitch from Robert Warner and Benjamin S. Williams, *The Orchid Album*, vol. 1 (1882).

benign manifestation of biophilia is our desire to share those new domesticated landscapes, specifically our homes, with a favoured selection of plant species, chosen for pleasure and companionship. The house plant is part of our innate need for novelty, a flicker of green life in an increasingly homogeneous world.

The nomenclature of house plants is necessarily vague: they can be referred to as indoor plants, pot plants, foliage plants or house plants. The term 'house plant' was coined in 1952 by the English nurseryman Thomas Rochford, whose stock of ornamental plants for the house had previously been called 'green plants' or 'foliage plants'.[3] In the United Kingdom Rochford's name quickly became synonymous with high-quality house plants, and, with the writer of horticultural manuals Dr D. G. Hessayon and the house-plant fertilizer Baby Bio, he formed part of a holy trinity for several generations of house-plant growers.[4]

House plants are usually tropical or subtropical in origin (although in the UK the common ivy, *Hedera helix*, remains a favourite), rooted and potted, intended as adornments for the home, and often

The love of house plants is a shared human attribute, crossing cultures and budgets. Carl Larsson, *Flowers on the Windowsill*, from the series *A Home*, 1895, watercolour on paper.

Scorched *Monstera deliciosa* plant leaf: each house plant survives within a house-specific equilibrium of care and neglect.

purchased to endorse an individual's cultural and social identity. A house plant in London may be a garden plant in Spain or Florida. In the tropics you have the strange situation where a tropical monstera may grace an air-conditioned living room while outside the window the same species is scrambling up a tree as a robust fruiting vine. There is a shady and flexible line between the house plant and the conservatory plant, and a narrow gradient between the house plant as benign decoration and the house plant as focus for obsession. There is also a fine thread between cut flowers and house plants; a pot of poinsettia, *Euphorbia pulcherrima*, bought for Christmas is in reality a rooted bunch of cuttings that will likely be thrown away after a few months. Similarly, the bound and braided rooted cuttings of *Dracaena sanderiana* that are sold in containers often without soil

Lucky bamboo, neither lucky nor bamboo, are cuttings of the central African *Dracaena sanderiana*.

(ironically known as lucky bamboo) rarely survive long. For this book the definition has been kept loose and flexible, allowing for a reasonable amount of meandering.

The culture of house plants varies across the globe, influenced by wealth, the available range of plants, culture and the prevailing sense of design.[5] Examples characteristic of their locations are Madagascar periwinkle, *Catharanthus roseus*, growing in tins decorating the entrances of homes in South Sudan; potted house plants, often bizarre succulents, crowding the front doors and windows of a Tokyo backstreet; and balconies packed with house plants and prized pots of mint or basil in Mediterranean cities such as Alexandria and Barcelona.

As house plants have moved around the globe they have been modified by each successive culture, with different regions developing favourite species and cultivars. For instance, the orange-flowered *Clivia* lilies from South Africa, domesticated and cultivated in Europe and North America, have become popular in China, Japan and Korea, with each country favouring different types of cultivar.[6] Different regions may have differing favourites. In the United States bigger

John Nash, *Window Plants*, 1940s, coloured woodcut. House plants are a long-established addition to the household, providing companionship and a bridge between home and a distant natural world.

Clivia 'Longwood Debutante', produced by the plant-breeding programme at Longwood Gardens, Pennsylvania.

plants such as *Dracaena* are popular, perhaps reflecting bigger rooms or the availability of large plants from Florida nurseries. In the United Kingdom and United States for some households the well-being of the house plants clearly takes priority over the comfort of the human residents.

Yet behind these variations there is now an increasingly cosmopolitan house-plant flora. Visit a shop in Beirut, London or Bogotá and you will find many of the same species, often the same cultivar; indeed, they may originate from the same growers or have arrived via the same horticultural auction house. The local cultures of house plants at a time when specific market-garden districts served a metropolitan area have been replaced by distant nurseries connected by aeroplanes, container shipments and long-distance haulage. This has

Roger Eliot Fry, *Clivia*, 1917, oil on canvas.

sadly suppressed the viability and creativity of small nurseries and independent breeders.

The public interest in house plants has increased dramatically in recent years. Visit a well-stocked house-plant shop and you will witness a phenomenon long thought impossible by many horticulturists

Garden House, Pali Hill, Mumbai, constructed by architects Disney Davis and Nitin Barcha, which demonstrates the link between plants and architecture and the desire for a more natural living space.

Plant shop in Melbourne. House plants are a huge international and multi-million-dollar trade connecting markets with growers in Thailand, China, the Netherlands, Australia and the United States.

— that of young people getting excited about buying plants.[7] As generations have moved along, so the memories of spider plants and macramé potholders have faded only to be revived by later generations as they have rediscovered the pleasures of growing plants. We are told that one in three Australian households has several house plants.[8] Plant sales have increased, numerous books on the subject have been published and large numbers of growers keep connected through social media; for instance, the House Plant Hobbyist Facebook group has over 366,000 members and the House Plant Growers group over 157,000.

This love of house plants is a contemporary riff on an ancient tradition.[9] Over thousands of years we have used plants, usually flowers and foliage, to mark the seasons and holy days, to bring luck and blessings to houses and to honour the choreography of life.[10] Quite probably this relationship dates back to prehistoric burials, and it flourishes today as the billion-dollar cut-flower market.[11] Many ancient societies used living plants to decorate temple complexes and palaces; one such example is Pharaoh Hatshepsut's (*c.* 1507–1458 BCE) expeditions to the distant Land of Punt (probably Somaliland) for incense trees, *Boswellia* spp., to grow in her temple complex.[12] This is not only an ancient case study in organized plant collecting; it demonstrates the pervasiveness of the cult of the exotic, a cult that still drives our horticultural passions. The word 'exotic' is ultimately derived from the Greek *exo* (outside), and refers to those artefacts or products from another place or culture.[13] It has connotations of the strange, the alluring, sometimes sinister, sometimes whimsical and, importantly with regards to house plants, tropical. For many of us, house plants are an affordable piece of exotica.

As jaded twenty-first-century citizens it is difficult for us to imagine the extraordinary impact of the first live tropical plants to have been exhibited in Elizabethan London. Then, the possession of exotica endowed the owner with an air of sophistication and worldly experience, demonstrating access to the rare, marvellous and expensive. Today, while some house plants can be exclusive and rare,

Ti plant, *Cordyline fruticosa*, an ancient Polynesian domesticated plant now widely cultivated as a tropical foliage and house plant. In Hawaii it is traditionally planted at doorways or house corners to bring luck to the household.

House plants represent a peculiar subset of the tropical plant kingdom that can be propagated at scale and survive the conditions of the house.

more often they are affordable, an impulse buy. Some are traded through non-cash transactions of cuttings and seeds, trades that contribute to the plant's role as a repository of personal history. Some we keep for decades, others are transient decorations to be thrown away, the dismal fate of the leafless poinsettia and rotted cyclamen in mid-January.

In Western European culture, we can trace the beginnings of a house-plant culture back to medieval times with the cultivation of gillyflowers, *Dianthus caryophyllus*. Valued for their beauty and scent, they were brought into the house for winter shelter.[14] However, it is not until the early seventeenth century that we have evidence of a range of plants being brought into the house for decoration, the 'garden within doors' described by Sir Hugh Platt (1552–1608) in his gardening manual *Floraes Paradise*: it is a 'pleasing thing to have a faire gallery, great chamber, or other lodging . . . to be inwardly garnished with sweet herbs and flowers, yea and fruit if it were possible'.[15] Later in the seventeenth century the traveller and diarist Celia

overleaf: The dark side of the house-plant trade: cacti dyed, denatured and dipped in wax.

Fiennes (1662–1741) observed that a number of exotic plants were kept within the house of the earls of Bedford at Woburn: 'just by the dining room window is all sorts of pots of flowers and curious greens, fine orange, citron and lemon trees and myrtles, striped filleroy [*Phillyrea*] and fine aloe plants.'[16] Thomas Fairchild (*c.* 1667–1729) published *The City Gardener* in 1722, reflecting the new interest of Londoners in 'furnishing their rooms or chambers with basons of flowers and bough-pots'.[17]

The great nineteenth-century fascination with house plants can be traced to a unique set of converging factors. The invention of the Wardian Case allowed plants to be moved from the tropics to the temperate world, and there was an increasingly affluent set of households (not only the aristocratic collector) who saw botany as a wholesome and rewarding hobby, and a set of tropical plant nurseries whose sole job was to tempt the hobbyist and collector. Many of the plants we grow today started the process of their domestication in the nineteenth-century botanic nurseries. Importantly, horticulture was being increasingly driven by a novel new technique, hybridization, that was adopted with gusto by Victorian nurseries.

The scale of the contemporary house-plant industry is staggering. In 2014 over 50 million poinsettia plants and 4.5 million African

Crowd of house-plant buyers at a flower market in Milan, 1979.

violets were sold in the United States, and the total value of the U.S. indoor foliage plant market was $747 million (about £559 million).[18] Individual plants can sell for large sums; in New Zealand in 2020 a single hoya plant was sold for NZ$6,500 (about £3,340), beating the previous record held by a monstera that sold for NZ$5,000 (about £2,570).[19]

The growing and transport of house plants is today a complex industry that links our homes and offices to distant growers in the Netherlands, Florida, Korea, Thailand and Costa Rica. The industry continues to balance traditional horticultural techniques, including the keen eye and vision of the accomplished breeder, with the high-tech tools of genetics and plant physiology to provide a steady succession of novelties.

As wealth and leisure time have increased, so our houses have become more welcoming to plants. Today, at least in the affluent corners of the world, our houses are better lit by larger windows, and heating is dependable and steady and, importantly, no longer provided by polluting coal fires and toxic gas. At one time the choice of plants was decided by what could survive toxic growing conditions. The aspidistra was one such survivor, as a Prohibition-era writer recorded: 'cases are known of this plant growing in the alcohol-laden air of an old saloon or in more modern speakeasies and beer gardens, receiving its only moisture in the form of discarded beer or dishwater.'[20]

We have banished a fear of house plants as a danger to our health, a source of toxic effluvia that could harm sleeping victims.[21] John Mollison in *The New Practical Window Gardener* (1877) warns that the 'blossoms of plants give out more carbon than any other part, therefore hand or table bouquets should not stand in your room during the hours of sleep.'[22] This is in contrast to today, when plants are widely promoted and mythologized as heroic mechanisms for filtering pollutants from indoor air.

Most of us live in crowded urban spaces. The urban biome covers about 3 per cent of the Earth's land surface and houses 55 per cent of its population. It is the planet's newest ecosystem and one where novel ecologies are developing.[23] A key part of the new urban ecosystem is the ecology of our indoor living spaces, our homes.[24] In cities access to nature is increasingly difficult and the need for creating a self-defined living space is growing stronger. House plants will continue to represent a large part of our daily dose of nature and contribute to our health and well-being through providing an outlet for our emotions and creativity. Innovative molecular research is showing that house plants are a dynamic part of the house ecology, interacting with and influencing the microbial diversity of a room.[25]

House plants are commodities: they are bought to bring pleasure, to define the individual household and sometimes to signal social prestige. They provide a focus for attention and emotional recharge. They provide the exhibits for our private cabinets of curiosities. House plants can share our homes for years, sometimes decades; indeed, some may last longer than a spouse and some will stay in a family over generations. Bought, exchanged, sometimes stolen, and given as gifts by millions of people, their fundamental purpose is to add pleasure and richness to life.

Those we grow today are derived from the flotsam and jetsam of centuries of plant collecting, and consist of a selection of plants that are both physiologically robust and attractive to the homeowner. These eclectic harvests from the tropical world carry a history that has often been lost, their origins clouded by the erasing forces of marketing and advertising. Indeed, some are sold with no name as one specimen in a batch labelled 'mixed cacti and succulents'. One UK supplier, Patch, has dropped the use of scientific names and rebranded its plants as Fidel (*Ficus lyrata*), Dora (*Alocasia* 'Portodora'), Mick (*Dracaena fragrans*) and Phil (*Philodendron scandens*), confirming their identity as commodities purged of ecological, taxonomic or historical context. One example, sold as Cassie – or as part of the 'Unkillable Set' – is *Zamioculcas zamiifolia*, collected in East Africa in 1869 by John Kirk, a colonial governor of Zanzibar and a major player in stopping the Indian Ocean slave trade. Traded globally under the name ZZ, it is regarded as a near-indestructible house plant, propagated by the million in Thailand and China, and sold without acknowledgement of ecology and history.

Like all biological or cultural collections, our houses contain intertwined stories of magic and misery. Some species are long-term house guests with centuries of coexistence in our homes, and carry the distant whiff of colonial enterprise. We know the India rubber plant, *Ficus elastica*, as a gangly house plant, but for the nineteenth-century economic botanist it was the source of a waterproof gum; similarly, the snake plant, *Sansevieria*, was planted as a fibre crop for the dry tropics.

Some of the early hunting grounds for plants were also the front lines for colonial military campaigners and traders. The South African partridge-breasted aloe, *Aloe variegata* (*Gonialoe variegata*), the 'gateway' succulent that can introduce susceptible innocents to the addictive world of cacti and succulent collecting, has been in cultivation in Europe since the 1680s, when it was collected from the Cape colony. Similarly, the dragon trees, *Dracaena*, were brought out of West Africa to Dutch collections in the 1690s and have been popular house plants for around 150 years. The now ubiquitous African violet (*Saintpaulia* spp.) was collected from the montane forests of Tanzania by a German colonial administrator in the early twentieth century and has been subject to more than a century of intense breeding to create a bewildering array of new cultivars (or abominations, depending on your point of view).

Some house plants resonate with a troubled history. *Dieffenbachia*, originating in the Caribbean and South America, is a storied plant and a long-cultivated ornamental aroid with wild species being hybridized as far back as the 1870s. Pre-Columbian cultures used it as a potent medicine and poison (for instance mixed with curare to tip arrows), but this potency was abused when the slave economy became established in the Caribbean, and the 'dumb-cane' was used as a brutal punishment for enslaved peoples.[26] The toxic power of this plant later led to proposals that it could be used by Nazi authorities for the mass sterilization of prisoners deemed racially inferior.[27]

Heidi Norton, *My Dieffenbachia Plant with Tarp*, 2011, archival pigment print.

One plant with many stories: a beautiful plant bred for commerce; a valued ethnobotanical resource for Pre-Columbian peoples; an agent for brutal racist punishment, while in the Brazilian Amazon it is grown on verandas as a magical defence for households against evil.[28]

We can conveniently attribute the start of Western house-plant culture to Hugh Platt's gardening manual *Floraes Paradise* (1608), and accordingly we have a four-hundred-year history of house plants. During this time, a cascade of species has been introduced and sold for house decoration, with many species coming in and out of favour as taste changes and methods of horticultural production favour the mass-produced, and mass-marketed. Some species have faded away after periods of horticultural fanaticism (for example ferns following 'Pteridomania' during the nineteenth century), and others are incredible survivors that have been through several cycles of love and abandonment – such as the Swiss cheese plant.

While house plants may be viewed as chic, naff, exotic, mundane or monstrous, they have played a role in the world of fashion and design, both stimulating and reflecting design movements. Today house plants are featured on clothes, fabrics, household goods and computer cases. In the early twentieth century they inspired the design of an Art Deco imperial railway station in Vienna, while Walter Gropius, founder of the Bauhaus school, was an inveterate cactus collector.

Tropical indoor plants have provided literature with a wide harvest of symbols and metaphors, and some plants have attracted particular attention in this regard.[29] For instance, *Caladium* or elephant ears were regarded by French *fin de siècle* writers as deeply tainted

Dieffenbachia, a plant with a powerful history as ornamental, ethnobotanical, cruel colonial poison and guardian of Amazonian households against bad luck.

House plants are part of interior design, often used as a statement of ambition and status, as shown in this 1938 photograph published in *Die Dame*.

symbols of decay. The house plant as killer has been explored in H. G. Wells's short story 'The Flowering of the Strange Orchid' (1894), and in films, among them the comedy musical *Little Shop of Horrors* (1960 and 1986), the science fiction *Invasion of the Body Snatchers* (1956 and 1978)[30] and the recent sci-fi drama *Little Joe* (2019). House plants have been celebrated in song – 'The Biggest Aspidistra in the World' by Gracie Fields – as a literary metaphor by Baudelaire and Émile Zola, and in fabric, sculpture and painting. Big-leafed exotica, including a recurrent motif of *Sansevieria*, appear in the jungle scenes of Henri 'Douanier' Rousseau, and are cheekily echoed in the backdrops of the animated film *Madagascar* (2005) by Dreamworks. An expanding range

Plants like *Sansevieria* have provided exotic motifs for art and design: Henri Rousseau, *The Equatorial Jungle*, 1909, oil on canvas.

of books, blogs and websites reflect an extraordinary interest in house plants. A look at the range of books available suggests that they fulfil a wide array of emotional needs (as perhaps exemplified by titles such as *How to Make a Plant Love You* or the calendar *Boys with Plants: Sexy Men Caring for Indoor Plants*). You can even listen to electronic music claiming to promote the health of house plants while selecting your new plant based on your horoscope.

The link between plants and people can be surprisingly intimate: Ramma Damma, aka Ulli Hopper, 'married' a potted plant (Xanthippe) at Gretna Green, Scotland, 2012.

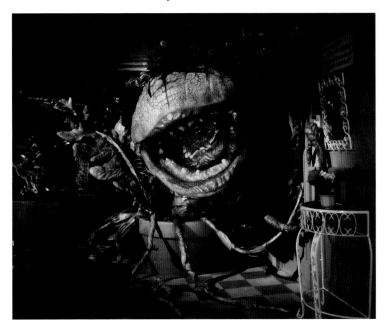

There is a dark side to the plants we have invited into our homes: the carnivorous Audrey II plant in *Little Shop of Horrors* (1986, dir. Frank Oz).

The breeding of house plants over the last four hundred years is a narrative of increasingly science-based approaches, from Thomas Fairchild's 'mule', the first garden hybrid, in 1717, to the application of genetic engineering for today's plants. It is also the evolution of an art form, each breeder employing an alchemical process that blends a personal vision for a plant with an understanding of the commercial market.

It has never been enough just to have a plant on a windowsill. As the interest in house plants has grown, so has the amount of equipment associated with their cultivation. In the nineteenth century the success of Nathaniel Ward's glass boxes allowed homeowners to grow a wide range of plants protected from toxic coal fumes.[31] Ward created a tool that allowed the shipping of delicate exotics across the globe, and as such both fuelled colonial industry and created a decorative centrepiece for Victorian homes. Today new horticultural tools allow us to grow spectacular green walls as inspired by the

AND INCORRECT FEEDING CAN
AFFECT THEM ADVERSELY

'And incorrect feeding can affect them adversely', cartoon
by Norman Thelwell, *Up the Garden Path* (1967).

botanist Patrick Blanc, and to build twenty-first-century versions of
the Wardian Case: terrariums and aquariums, as exemplified by the
beautiful aquatic installations by the late Takashi Amano.[32]

House plants enter our homes through a variety of routes, and we
have a different set of relationships with each one. You could argue
that each collection of house plants represents a personal cabinet of
curiosity, a *theatrum memoriae*, meshed together by stories, memories
and our individual sense of curatorship. Each collection provides a
mix of continuity and change, as the memories carried by each plant
are enhanced by the anticipation of new growth or flowering.

Some growers have a relaxed relationship with their plants ('they
have two chances, die or thrive') and are happy enough with those
plants that are tough enough to survive the house-specific equilibrium

of love and neglect. At the other end of the spectrum is the specialist collector, who chases down the rarest and finest plants, ensures he or she is at the head of the queue for specialist plant-society sales, is on first-name terms with the best growers, and talks with extraordinary enthusiasm about a new reverse-osmosis watering system. This individual develops a deep, often unique expertise at growing and places a dubious value on specimens with a collection number, the shamanic incantation linking that plant to a point of collection from the wild. Wild collected plants as a rule should not be purchased.

Gilding the cactus: artificial flowers on potted cacti.

It is becoming increasingly difficult to differentiate between the living and the dead: a selection of plastic plant arrangements.

Somewhere between them is the individual who grows a wide diversity of house plants from a range of sources and is less focused on the apparatus while generous with horticultural hints and cuttings. This happy accumulator takes time to learn about their plants but is fatalistic about loss, ecstatic about success, and subject to those powerful waves of enthusiasm whereby different plant groups are favoured and abandoned through successive two-to-five-year cycles.

There is a dark and melancholy side to the cultivation of house plants: a dismal area where plants are dipped in wax, dusted in glitter, dyed with fluorescent colours and pinned with fake flowers; a shadowland of hurried impulse buys, disappointment and slow vegetal death. This in turn merges, increasingly easily, into the dark kingdom of the artificial house plant, where now it is becoming difficult to separate the living and the dead.[33]

There is a long and ignoble tradition of stealing house plants, usually the pinching of cuttings. Stories abound of coachloads of garden societies descending on public gardens and leaving with bags

A NIGHT BLOOMING CEREUS

COBEA VINE IN A WINDOW

Over the last century our expertise in house-plant horticulture has continued to improve: photographs of a night-blooming cereus (left) and a cobea vine (right), from Parker T. Barnes, *House Plants and How to Grow Them* (1909).

stuffed with pilfered stems. The real trouble occurs when the stolen plant is slow-growing, valuable or threatened with extinction. The impact of the theft on the grower can be emotionally and financially devastating. A case in point was the theft from the Royal Botanic Gardens, Kew, of the dwarf water lily *Nymphaea thermarum*, which survived only in cultivation after the loss of its sole wild habitat in Rwanda.[34]

Now is a good time to review the history and function of house plants. There is undoubtedly a house-plant renaissance, with an ever-increasing range of species being modified and adopted for domestic cultivation. The dynastic house-plant innovators, such as Veitch, Loddiges, van Houtte and Rochford, would look in wonder at the range of plants we grow today, their quality and the cutting-edge science that supports this global trade. They would be mystified by the social-media 'plantfluencers' but inspired by the online communities exchanging information on how to grow house plants.

Over the four hundred years since the publication of *Floraes Paradise*, the house plant has evolved from a minor commercial crop, often propagated and sold locally, into a global one. It has diversified from a small number of species to encompass hundreds of species with thousands of cultivars – perhaps the most eclectic experiment in plant domestication on Earth. The diversification of house plants has been driven by evolving horticultural technology and will increasingly be moulded by the new molecular tools. Perhaps most exciting is the evolution of the house plant from a specimen to a landscape, with green walls and towers wrapped inside and out by plants. This is the botany of the urban biome.

The New Guinea Impatiens hybrids were bred from wild collected plants as part of a Longwood Gardens breeding programme.

one

The Gathering of the Exotic

🪷

'The greatest service which can be rendered any
country is to add a useful plant to its culture.'

THOMAS JEFFERSON, 3rd president of the United States, *c.* 1800

A visit to a good house-plant shop or garden centre reveals a range of plants with an astonishing diversity of shapes and colours: the African violets from east Africa with blue and pink flowers; the deep green, heraldic leaves of *Alocasia* from Asia; the geometric rosettes, sometimes more like a lurid cauliflower, of the Mexican succulents *Echeveria*; the scarlets and pinks of the *Poinsettia* leaf bracts from Mexico; and the tall, pale candelabrum of *Euphorbia ingens* from southern Africa. Collectively this selection of plants spans a set of wild habitats, respectively, that include the misty upland forests of Tanzania, Asian rainforests, the dry deciduous woodlands of Mexico and semi-arid southern Africa. Each plant has a story, starting from the point of collection in the wild and encompassing the subsequent saga of breeding and selection over many decades (in some cases, centuries). From the moment the plant or seed was collected from the wild, these plants have been moulded into human artefacts often increasingly removed in shape, colour and physiology from the wild stock.

House plants have arrived in Western cultivation by a variety of routes, some via scientific expeditions, such as Isaac Bayley Balfour's nineteenth-century expedition to Socotra, bringing back *Begonia socotrana* and *Exacum affine*; others via the diplomatic bag, such as the first African violets from colonial Tanzania; and yet others through commercial collectors, such as the 'travellers' sent out by the Veitch nurseries. Sometimes they have been released, or leaked, from private collections or university gardens. In a few cases there have been dedicated expeditions to bring back potential house plants for breeding, such as the Longwood Gardens expeditions to New Guinea for *Impatiens* in 1970.[1] However, the most useful prospecting fields for new plants will continue to be other people's collections and nurseries; the hard fieldwork has been done, a keen eye has already scanned the stock and with luck the growing needs of the plant are understood.

Many a plant breeder has seen the potential for transforming an oddity or curiosity into a commercial crop, an undertaking that can involve large financial investment and decades of research. It is a strange alchemical process, between science and art, increasingly driven by molecular genetics and yet still strongly influenced by the breeder's intuition. For most species it is a complex progression of stop and start, as different breeders carry the flame and as fashions come and go.

The origin of the African mask plant, *Alocasia* × *amazonica*, illustrates the peculiar pathways to domestication for some house plants. This beautiful aroid, its leaves deep green with white veins and often a deep purple underside, is an artificial hybrid between two Asian species, the large-leafed *A. watsoniana* (part of the *A. longiloba* complex) and *A. sanderiana*, named for the UK's great nineteenth-century orchid nursery Sander's of St Albans and endemic to the Philippines, where it is gravely threatened with extinction.[2] The 'Amazon' of the cultivar name commemorates the nursery in Miami where the hybrid was made by plant breeder Salvadore Mauro in the 1950s.[3] While this artificial hybrid is neither African nor Amazonian, it is splendidly exotic.

When the garden writer Hugh Platt published his *Floraes Paradise*, in 1608, there was no grand view of the plant diversity of the tropics.[4] The region was poorly understood, there was no accessible library of images to inform and tempt the northern horticulturist, it was virtually impossible to transport live tropical plants and there was little or no understanding of how to cultivate them. However, exotics slowly started appearing in European markets; for instance, the royal apothecary and pioneering cactus collector Hugh Morgan (1530–1613) grew, and probably sold, Caribbean cacti (*Melocactus*) in London.[5] The pioneering field botanist and apothecary Thomas Johnson displayed the first bunch of bananas in his Holborn shop, in the City of London, in 1633.[6] It is now difficult to imagine the extraordinary impact of these strange plants on a population whose experience of plants was still fundamentally European, when the remote and dangerous tropics were far beyond the imagination of botanists and collectors.

As global trade developed, so the idea that the tropics held previously unsuspected diversity and botanical splendour became established in the Western consciousness. The use of the term 'exotic' for uncommon and cold tender plants can be traced to the English herbalist John Gerard, who used the term 'exotick' in his *Herball* (1597).[7] A number of exotic plants were introduced from South America at the time that are still stalwarts of the summer or tropical garden, including the tuberose (*Polianthes tuberosa*, now *Agave amica*) and the marvel of Peru (*Mirabilis jalapa*), both gloriously scented.

These new and tender plants arrived without instruction books, and talented horticulturists, such as Hugh Morgan and John Tradescant the Elder (*c.* 1570–1638), must have used their well-honed intuition to work out how to grow and propagate such species. We can guess that many of these new exotics would have been stored for the winter in sheds, probably wrapped in straw, with some being brought into the house. Winter mortality would have been high. Later, John Evelyn (1620–1706) in his unpublished *Elysium Britannicum* describes an early glasshouse being heated by 'a large pan of coales thoroughly kindled . . . and then placing it upon a hand barrow, have two men carrie it gently about the conservatory'.[8] It is easy to imagine the dreadful condition of both plants and the horticulturists' lungs by the end of a hard winter.

The big leaves of the tropical aroids, in this case *Philodendron giganteum*, have fascinated horticulturists for centuries.

The tropical aroids, and in particular the New World large-leafed philodendrons and monstera, provide an illustration of how botanical discovery and commercial description fuelled horticultural passion and the subsequent adoption of these species as popular house plants.[9] Since their earliest introduction northern horticulturists have been fascinated by these tantalizing fragments of a distant 'torrid zone'. The aroid plant family, Araceae, supplies some of the most popular and spectacular house plants. These include the Chinese evergreen, *Aglaonema*; the large-leafed *Alocasia* and *Colocasia*; the flamingo-flowered *Anthurium*; the garish but splendid *Caladium*; the spot-leafed dumb cane, *Dieffenbachia*; the rampant climbers *Epipremnum*, *Monstera*, *Scindapsus* and *Philodendron*; the cycad lookalike *Zamioculcas*; and the white-flowered peace lily *Spathiphyllum*. These are predominantly tropical forest plants that largely grow as semi- or full epiphytes. *Zamioculcas*, in contrast, is a terrestrial plant of dry bushland, and – uniquely in the Araceae – uses the CAM (Crassulacean acid metabolism) photosynthetic pathway, a water-efficient variation of photosynthesis that allows the plant to grow in seasonally dry habitat, or to survive as a neglected house plant.[10] There is another set of tropical aroids that traditionally are cultivated in tropical fish tanks but are increasingly appearing in terraria (or paludaria), including *Cryptocoryne* from Asia and *Anubias* from Africa.

The temperate world's fascination with tropical aroids started with the French botanist and priest Charles Plumier (1646–1704) collecting in the Caribbean. While tropical aroids (*Philodendron* spp.) had been collected earlier in Brazil by the German naturalist Georg Marcgrave (1610–1644), their scientific study really began with Plumier, who published an account of his trip of 1689 (*Description des plantes de l'Amerique*, 1693). In it he illustrated the strange growth and spectacular leaves of these plants. He made a second voyage in 1693 and a third in 1695. Like so many naturalists of the time, Plumier described and portrayed species that could neither be cultivated in nor transported live to Europe; these extraordinary plants remained unattainable and intriguing objects of curiosity.

A series of great botanists followed Plumier, each revealing and cataloguing the extraordinary abundance of the tropical aroids. Nikolaus Joseph von Jacquin (1727–1817), one-time director of the

The dissected leaves and aerial roots of *Philodendron mello-barretoanum* make a distinctive and large house plant.

41

Zamioculcas zamiifolia, from the arid lands of Eastern Africa and known today as the zz, sold as a virtually indestructible house plant. Illustration from *Curtis's Botanical Magazine*, vol. xcviii (1872).

The genus *Aglaonema*, the Chinese evergreen, originates from Southeast Asia and New Guinea and is one of the more popular aroid house plants.

University of Vienna botanic garden, collected plants and animals for the Imperial gardens at Schönbrunn, Vienna. He had specific but perhaps not entirely helpful instructions about collecting priorities from his employer, Emperor Francis I: 'lions or tigers are to be excluded from this mission,' allowing him to 'personally choose those species of flowers that are rare and deserve to be in my garden . . . only those that are beautiful or that have a pleasant scent . . . he must, however, restrain somewhat the inclinations of the gardener, who may believe that everything that is available is good.'[1] Von Jacquin spent four years working in the Caribbean (1755–9) and there collected tropical aroids, including *Philodendron*.

Following von Jacquin was a fellow Austrian, the botanist Heinrich Wilhelm Schott (1794–1865), the son of a horticulturist at the University of Vienna botanic garden, where he grew up among rich plant collections. In 1815 he became a horticulturist at the Belvedere Palace, and then at von Jacquin's recommendation he was invited to join a scientific expedition to Brazil with the botanists Carl Friedrich Philipp von Martius, Johann Baptist von Spix and Johann Christian Mikan. On Schott's return he was awarded the directorship of the botanical and zoological collections of the Imperial Palace of Schönbrunn. He was a man of extraordinary energy and scientific productivity. He described many of the tropical genera now popular as house plants, including *Anthurium, Pothos, Epipremnum, Monstera, Spathiphyllum, Philodendron* and *Dieffenbachia* (the last of which he named after Joseph Dieffenbach, head gardener at Schönbrunn). Schott was the first monographer of the Araceae, and perhaps his greatest legacy is the series of over 3,400 magnificent watercolour and pencil drawings of spectacular leafed aroids contained in his book *Icones Aroidearum* (1857). While Plumier's line drawings introduced the bizarre leaf forms of the aroids, it was Schott who portrayed the dazzling colours and textures of these plants. The watercolours were based on plants grown at Schönbrunn, and this extraordinary collection of images probably started the horticultural craze for the tropical aroids. It was the frontispiece to Schott's monumental *Aroideae Maximilianae* (1879) that crystallized an increasingly enchanting vision of the aroids as part of a benign and abundant tropics. It shows great tropical trees wrapped with tropical aroids while those other tropical icons, macaws and blue morpho butterflies, flit through the forest.

The plant that perhaps exemplifies these tropical aroids is the spectacular, and perhaps overly familiar, *Monstera deliciosa*, the Swiss cheese plant.[12] Charles Plumier was the first Western botanist to describe and illustrate the genus *Monstera* under the name *Arum hederaceum amplis foliis perforates*. The species you will find in your plant shop today was first collected in Mexico in 1832 by the Hungarian botanist Wilhelm Friedrich Karwinsky von Karwin (1780–1855). However, his dried specimen was lost in the Munich herbarium. Karwinsky has other claims to fame; the charming but pernicious *Erigeron karvinskianus* is named after him, and he also sent living material of the poinsettia from Mexico to Germany.

M. deliciosa was next collected in 1842 by the Danish botanist Frederik Liebmann (1813–1856), who named the species and brought

The leaves of *Monstera deliciosa*, a spectacular house plant and rainforest climber, and an iconic big-leafed exotic.

cuttings from Mexico to Copenhagen. After that a collection was made in 1846 by the Polish botanist Józef Warszewicz Ritter von Rawicz (1812–1866), who sent cuttings to Berlin from Guatemala. It is quite likely that the plants gracing front rooms today are derived from the two collections by Liebmann and Warszewicz. M. *deliciosa* is now a member of the pan-tropical garden flora and to many people a splendid symbol of the big-leafed tropics. To others it is fundamentally kitsch, part of the pantheon of tropical 'tat' along with plastic flamingos, tiki bars and overblown hibiscus flowers.

Viewing a wild monstera reveals that, as with many other house plants, we grow a stunted shadow of the real thing. In the wild, or in a tropical garden, monstera makes a spectacular vine that climbs up the host tree towards the light, reaching 20 metres (66 ft) or more. Only when mature does it flower and fruit; the latter is beautifully fragrant and tastes, when ripe, somewhere between a pineapple and a guava. The vine will start life as a seedling, usually on the ground, and grows horizontally as a very straggly, tiny-leafed vine.[13] Here it exhibits strange behaviour: it grows towards the darkest area of the tree canopy (a growth pattern called skototropism), and only when it has found a tree and attached itself does it start growing up towards the light.[14] As it grows upwards into better light the leaves get bigger, the stem thicker and the perforations more marked. All the monstera plants we grow as house plants are physiologically adult, having the perforated leaves associated with full sun, and are propagated by cuttings from plants with mature foliage.

Tyler Whittle, historian of plant hunting, divided the history of plant introductions into the pre- and post-Wardian Case periods, recognizing that the Wardian Case opened the door to the safe transportation of exotic plants to and from Europe.[15] As we have seen, it also allowed the successful cultivation of tropical plants inside polluted Victorian houses. During this period there was an extraordinary growth of expertise in tropical plant horticulture, with the great commercial nurseries anxious to satisfy, and encourage, the passions of the collector, whether individual grower or grand estate. Imported plants became envoys of distant worlds. They represented many things: an evocative whiff of adventure, romance and escape; commercial units of trade and profit; palettes for the ambitious plant breeder; and above all indicators of status and ambition.

The co-evolution of tropical house plant and grower had started. The nineteenth-century nurserymen employed collectors who engaged in ferociously competitive field collecting for the next great aroid to boost sales and prestige for a business fundamentally based on novelty. One such company was the Veitch Nurseries of London and Exeter, the largest family-run nursery in Europe. Veitch ran a team of plant collectors, and many of them made important aroid introductions that have become popular house plants. These include collections of *Alocasia* and *Scindapsus* species from tropical Asia by Thomas Lobb and John Gould Veitch; *Aglaonema* species, also from tropical Asia, by Charles Curtis; *Anthurium* species from South America by A. R. Endres, Gustav Wallis and Guillermo Kalbreyer (including the spectacular *A. veitchii*); and *Dieffenbachia* species from South America by David Bowman and Richard Pearce. The Veitch Nursery expected total loyalty and commercial results from their collectors; they were fulsome in their praise of the successful and damning of those who disappointed. In the *Hortus Veitchii* (1906), the published history of the company, one such collector, who is best kept anonymous, was described as having 'no special aptitude for collecting and entirely lacked the explorer's instinct . . . had to be recalled'.[16] Some Veitch 'travellers', such as William Lobb, Richard Pearce, David Bowman, Henry Hutton, Gottlieb Zahn, J. Henry Chesterton and Gustav Wallis, did not make it home, having died overseas.

Richard Steele in his *An Essay Upon Gardening* (1793) encouraged the gentleman collector to build hothouses and to take advantage of the new introductions, 'the prodigious varieties of rare plants that have been introduced' by 'men of greatest accomplishments [who have] navigated unknown seas, have traversed drear isles and deserts, searched the forests of both Indies and explored the burning countries of the torrid zone'.[17] The collector's zeal was further encouraged by

The rigid cycad-like leaves of *Zamioculcas*.

the number of magnificent new horticultural publications illustrating the latest plant introductions from the tropics. In some cases, the magazines were produced by nurserymen; the *Botanical Cabinet*, for example, was produced by Loddiges of London and the *Flore des serres et des jardins de l'Europe* by van Houtte of Belgium. Other beautifully illustrated journals include *Curtis's Botanical Magazine* (which is still in print), *L'Illustration horticole* and *La Belgique horticole*.

Some of the tropical aroids were 'sleepers', persisting in cultivation until fashion or technology released them into the house-plant industry. One such species is *Zamioculcas zamiifolia*, an aroid that looks like a cycad or succulent. ZZ, as it is called in the commercial trade,[18] was first described and illustrated in 1828 as *Caladium zamiaefolium*, but was moved to *Zamioculcas* by Schott. It was originally thought to have originated in Brazil, but plants were sent to Kew from Zanzibar by John Kirk in 1869. This is one of the few African aroids that are widely grown in Europe and North America, and it has developed a reputation as nearly indestructible – but it was in cultivation for more than a hundred years before commercial horticulture took an interest. The plant on the windowsill is virtually unchanged from the original wild plants, if fleshier and glossier than the drought-scarred wild specimens.

Other species caught the eye of the horticulturist almost immediately on their introduction and have been transformed by canny breeders over the centuries. One example is the aroid *Caladium* from tropical South America, first developed by European breeders in the nineteenth century, then by American breeders, and now reinvented by Thai breeders. *Caladium* was discovered in Brazil in 1767 by the French naturalist Philibert Commerson, and plants soon arrived in the United Kingdom. *Hortus Kewensis* listed it as *Arum bicolor* in 1789. The first Europeans to breed caladiums were the Belgian nurseryman

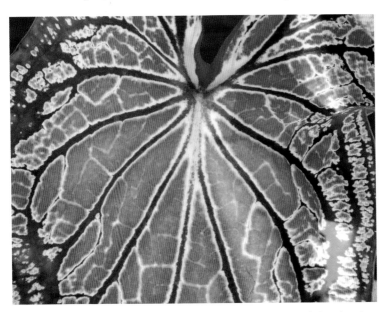

Close-up of the foliage from a Thai caladium hybrid. Over 150 years caladium breeding has moved from France to the United States, and now to Thailand.

Louis van Houtte and the French orchid hybridist Alfred Bleu in the 1860s; two of their hybrids, 'Triomphe de l'Exposition' and 'Candidum', remain in trade today.

The first public exhibition of caladium took place in 1867 at the Exposition Universelle in Paris. Caladiums were first introduced into the United States from Brazil in 1893 at the Chicago World's Fair, with a set of plants provided by the German-Brazilian horticulturist Adolph Leitze. The pioneering Florida nurseryman Henry Nehrling bought the plants and started breeding them at his Palm Cottage Gardens in Gotha, near Orlando. At one point he was reputed to be planting 250,000 tubers every year, representing around 1,500 cultivars. A number of these are still in trade today, including 'Mrs W. B. Halderman', 'Arno Nehrling', 'John Peed' and 'Fannie Munson'. The spectacular leaf colours and patterns of *Caladium* are one of horticulture's great 'Marmite tests': to some splendidly exotic, to others horribly lurid. The writers of *fin de siècle* France, such as Joris-Karl Huysmans, saw in their colours and textures the morbidity of flesh:

> The gardeners descended from their carts a collection of caladiums that pressed turgid and hairy stems of huge leaves, the shape of a heart; while retaining an air of kinship between them, none repeated themselves.
>
> There were extraordinary, pinkish ones, such as the Virginal, which seemed cut out of varnished canvas, in taffeta from England; all whites, such as Albane, which appeared to be cut from the transparent lip of an ox . . . these, like the Aurora Borealis, spread a leaf of raw meat, streaked with purple ribs, purple fibrils, a swollen leaf, sweating blue wine and blood.[19]

Historically, commercial caladium production has been focused on Lake Placid in central Florida, where there is approximately 485 hectares (1,200 ac) of caladium production. In summer these production fields have been likened to the bulb fields of Keukenhof

Recent breeding, particularly in Thailand, has transformed house plants such as the *Aglaonema*.

in the Netherlands – albeit with rather more mosquitoes. The next chapter of the caladium's development moves from Florida to Thailand, where a new generation of breeders have produced spectacular novel cultivars. Once the prerogative of the royal gardens in Thailand, the Thai caladiums are splendidly exuberant with glossier leaves and extraordinarily vivid colours, including yellows and deep, shiny scarlets. This is a magic the Thai breeders have worked on other tropical ornamentals, such as the desert rose (*Adenium*), *Aglaonema* and the Malagasy euphorbias, such as *E. milii*.

The poinsettia, *Euphorbia pulcherrima*, sits in a group of throwaway seasonal house plants, residents of the house for a few months only, and disposed of after the flowers are over. This is not a species for a long-term relationship. It is, however, one of the most important house plants in the global trade. The common name, poinsettia, honours the U.S. diplomat and amateur botanist Joel Roberts Poinsett (1779–1851), who found the plant in southern Mexico.[20] Poinsett was the first U.S. Minister Plenipotentiary in Mexico, and in this role travelled through the country in 1828 with colleagues from the Philadelphia scientific community. The party included William Maclure, a long-time friend of Poinsett; William Keating, a geologist; and Thomas Say, of the Bartram family of botanists.

It is thought most likely that Maclure brought the poinsettia back to Bartram's Botanic Garden in Philadelphia before Poinsett returned home. Colonel Robert Carr, who ran the botanic garden, recognized a star plant, and in June 1829 he entered the plant in the Pennsylvania Horticultural Society's very first flower show, as 'a new euphorbia with bright scarlet bracts or floral leaves, presented to the Bartram Collection by Mr Poinsett, United States Minister of Mexico'. From Philadelphia plants were shipped to the Royal Botanic Garden, Edinburgh. Poinsett's other great legacies are not always recognized. He was instrumental in setting up the United States Exploring Expedition (1838–42) and was a founding member of the National

Poinsettias, a classic Christmas plant for millions of homes, but very few survive past February.

NEW
CHRISTMAS POINSETTIA
SANTA

JOHN LEWIS CHILDS SEED CO. INC.
FLORAL PARK~NEW YORK
CONSOLIDATED WITH EDWARD T. BROMFIELD SEED CO.

Poinsettias on the back cover of John Lewis Childs's autumn seed catalogue, 1923.

Institute of Science and the Useful, known today as the Smithsonian Institution.

While Poinsett should be congratulated for spotting this plant, it is wrong to say he discovered it. The species was culturally important to the Aztecs, who called it *cuetlaxochitl* and used it as a dye, a cosmetic and to decorate royal palaces and temples. It was perhaps inevitable that such a spectacular plant would be observed by travelling colonial botanists; indeed the great and the good of early Mexican colonial botany all noticed the glorious red bracts.[21] The manuscripts of Francisco Hernández de Toledo (*c.* 1515–1587) mention the *cuetlax-ochitl* with 'very colourful tree leaves'. The Royal Botanical Expedition to New Spain (1787–1803), led by the renowned scientists Martín Sessé y Lacasta and José Mariano Mociño, collected the first scientific specimens and sent the first illustrations to Europe under the name *Euphorbia fastuosa*. Next in line, in 1803, Alexander von Humboldt and Aimé Bonpland arrived in Mexico and during their travels made collections of the plant under the names *E. coccinea* and *E. diversifolia*. Then followed Christian Julius Wilhelm Schiede and Ferdinand Deppe in 1828. About the same time that Poinsett made his collection, in 1833, Karwinski sent living material back to Berlin, and it was from this collection that Johann Klotzsch named the plant *E. pulcherrima*, the binomial we use today.

The species was absorbed into the Christian iconography of Mexico, becoming *la flore de Nochebuena* because it flowered around Christmas, a convenient bridging icon between the original Aztec ceremonials and the new Christian religious symbolism. As a result, Franciscan priests began to use the flower in Nativity processions.

In the wild poinsettias grow into a willowy shrub or small tree up to 4 metres (13 ft) tall. The house plant of today is a short version of the wild plant, suitable for indoor spaces and with bracts of varying colour (including a pointless white), intensity and texture. The domestication of the plant really started about a century after its introduction, when the horticulturist Paul Ecke Sr (1895–1991) took the baton. In the 1920s he saw the potential of this subtropical shrub as a house plant and started growing it in California, first in Hollywood and then in Encinitas, where the Paul Ecke Ranch continues to produce poinsettias. Through his marketing efforts the poinsettia has become symbolic of Christmas in the United States and beyond.

The domestication of the poinsettia has been marked by a series of technical breakthroughs that reflect the increasing sophistication of horticultural research and production.[22] The early plants were fragile, the colourful bracts short-lived and the leaves prone to dropping, but despite this they became popular house plants. One particular problem was the difficulty of scheduling the plants for Christmas sales. The discovery of photoperiodism, the reaction of plant growth to differing periods of night and day, led to the use

Poinsettias decorating the plaza outside the Metropolitan Cathedral, Mexico City.

of blackout curtains in nurseries to lengthen the night period and encourage synchronous flowering of the crop. From the mid-1950s there was a focus on selective breeding that resulted in improvements in colour and the poinsettias' robustness as a crop, notably delayed leaf ageing and reduced leaf drop. In the 1980s it was discovered that introducing benign phytoplasmas bacteria via grafts into plants increased branching and therefore resulted in more flowers. Other innovations included better pruning and the use of chemical growth regulators to effectively shrink the plants by reducing the length of the stems. More recently, genetic manipulation has been used to improve disease resistance.[23] The mode of propagation has also changed fundamentally. In the early twentieth century plants were field-grown in California, Florida and Texas and shipped bare-rooted as mother plants to produce cuttings for growing near the markets of the northeastern United States; today cuttings are flown in from South American nurseries.

A genetic analysis of wild poinsettia populations has confirmed the origin of the original Poinsett collections from the vicinity of Taxco in northern Guerrero, Mexico. This study also discovered that the existing breeding programmes have used only a small portion of the total genetic diversity of the species, meaning that a large amount of wild genetic diversity is available for incorporation into new poinsettia breeding programmes.[24] Here is an opportunity to lay to rest a long-standing animosity in Mexico towards the legacy of Joel Poinsett, where the adjective *poinsettismo* is used to describe U.S. arrogance and meddling in Mexican affairs. A new generation of poinsettia cultivars are being bred from Mexican plants by Mexican breeders, so the *cuetlaxochitl* has returned home.[25] In 2002 an Act of Congress in the United States set 12 December, the day of Poinsett's death, as National Poinsettia Day to commemorate the man and his plant.

As the exploration of the world by Western powers proceeded, so new areas of botanical discovery opened up. One such area was the Cape region of South Africa, one of the first frontiers where Western scientists realized the vast diversity of plants existing beyond the

Mediterranean and north Atlantic islands. It proved to be the origin of many valued house plants. By the early seventeenth century a variety of plants from the Cape could be found in Dutch gardens, and from there they spread through Europe. These early introductions included the ornamental bulbs *Ornithogalum, Haemanthus, Nerine, Zantedeschia* and *Amaryllis belladonna*. By the 1630s John Tradescant the Elder was growing the beautiful *Pelargonium triste*, and so began the British love affair with 'geraniums'. The South African 'geraniums' were assigned to the genus *Pelargonium* by Charles Louis L'Héritier de Brutelle in 1789, so

Rembrandt Peale, *Rubens Peale with a Geranium*, 1801, oil on canvas. In reality Rubens is holding a *Pelargonium*. Rubens and Rembrandt were sons of Charles Wilson Peale, who established the first scientific museum in the United States.

separating the hardy herbaceous *Geranium* from the tender *Pelargonium*. This caused a confusion that has persisted, as we see from Shirley Hibberd's *The Amateur's Greenhouse and Conservatory* (1873):

> If there are any botanists within hearing, say pelargonium, and take all the consequences. But if none of these exacting and fastidious gentry are in the field, speak of the plants as 'geraniums' and you will have the good fortune to be understood by the entire audience without exception.[26]

One famous house plant, the spider plant, *Chlorophytum comosum*, originates in southern Africa, where it was discovered by Carl Peter Thunberg in 1794. This may be the most widely cultivated house plant, and is perhaps among the least inspiring. The generic name, *Chlorophytum*, simply means 'green plant'. However, sometimes a plant will emerge into the market as a mystery, a surprise without history. One such plant is the spectacular C. 'Fire Flash', the orange spider plant (seldom have the words 'spectacular' and 'spider plant' been used in the same sentence), which with its glossy leaves and bright orange petioles and leaf midribs is becoming increasingly popular.

Chlorophytum 'Fire Flash', a spectacular relative of the more abundant spider plant.

The patterned leaves of *Caladium*, sitting somewhere between exotic and lurid.

Grown, and probably selected, in Thailand, it is derived from one of the African wild species.

Freesias, another gift of the Cape, have long been cultivated to bring colour and scent into the home.[27] The development of freesia as a commercial crop and house plant can be traced back to Maximilian Leichtlin (1831–1910), bulb expert and breeder, who in the 1870s started crossing some of the wild species including unnamed yellow-flowered plants he found growing in the botanic garden of Padua (he later named this stock *F. leichtlinii*). A red-flowered sport of *F. corymbosa* was used to introduce new colours, and by the early twentieth century Dutch nurseries, notably Van Tubergen, were marketing dozens of new cultivars.

Two characteristic Cape genera, geranium (*Pelargonium*) and the Cape heaths (*Erica*), became hugely popular as glasshouse ornamentals and pot plants in Europe during the eighteenth and nineteenth centuries. The craze for Cape heaths had fizzled out by the 1850s, but a few are still available as winter-flowering house plants from a dwindling number of nurserymen in the United Kingdom, part of the lost tribe of cool glasshouse plants grown for winter decoration that

Freesias, valued for their scent and colour, are part of the great horticultural heritage of South Africa.

include the winter cherry (*Solanum pseudocapsicum*), the kangaroo vine (*Cissus antarctica*) and *Primula × kewensis*.

The humble 'geranium' has kept its place of pride as house plant and hanging-basket resident.[28] The story starts with *Pelargonium triste* coming into Europe in the seventeenth century and later with *P. zonale* being grown by the Duchess of Beaufort in the early eighteenth. The widely grown ivy-leafed pelargonium, *P. peltatum*, had arrived from the Cape via Leiden by the early eighteenth century. By the 1820s pelargoniums were established as garden and pot plants, and James Colvill's London nursery stocked around five hundred cultivars and hybrids. Robert Sweet (1783–1835) used plants from Colvill's collection and is recognized as the pioneer hybridizer in the United Kingdom. The Regal pelargoniums resulted from a cross between *P. cucullatum* and *P. grandiflorum*, with other species being added to blend, for instance *P. fulgidum* and *P. betulinum*, by breeders such as William Bull of the King's Road in London.

Another long-standing house plant from South Africa is the gesneriad *Streptocarpus*. Recently this genus has been expanded to include the African violets.[29] The breeding of this lovely plant has involved a series of horticultural giants who over 150 years bred a deservedly popular house plant. The first species to be introduced into cultivation was from Knysna, South Africa, by Kew collector James Bowie in 1826. The first commercial hybrids were bred by the French nurseryman Victoire Lemoine and offered for sale in 1859. In the UK the nursery company of Veitch and the horticulturists at the Royal Botanic Gardens, Kew (particularly William Watson), all played with the influx of new species. Giants such as Joseph Hooker, director at Kew, fell in love with the genus. The hybridization work was continued by John Heal of Veitch; Heal was one of Veitch's best hybridists and was famed for his hybridization work with orchids, hippeastrums, winter-flowering begonias and tender rhododendrons. After Heal, the work continued with Richard Lynch of the Cambridge University Botanic Garden. In the 1930s W.J.C. Lawrence of the John Innes Horticultural Institute bred two important cultivars, *S.* 'Merton Giant' and *S.* 'Constant Nymph', the latter being the basis of many of today's wonderful cultivars. Later work at John Innes,

Originating from southern Africa, *Streptocarpus* has been subject to generations of hybridization involving numerous wild species. Illustration from *L'Illustration horticole*, vol. XXXVIII (1891).

Gloxinia

1) Madame Pescatore. 2) Lilacina striata. 3) Mme de Parpart
4) Baronne de Champy. 5) Handleyana striata. 6) Mme Clementine.

Lith. Anst. v. A. Kolb. Nürnbg

In contrast to many valued house plants where interspecific hybridization has been used to create new cultivars, the cultivated *gloxínia* (*Sinningia speciosa*) is derived from one wild population in southern Brazil. Illustration from *Gartenflora*, vol. 1 (January 1852).

using X-ray-induced mutations, created plants that would flower all year.[30] *Streptocarpus*, like many house plants, has been moulded by a succession of breeders, each passing on the flame, using their intuition and the best available science.

Another gesneriad, but one from South America, the gloxinia, has a very different history of collection and subsequent breeding. In contrast to *Streptocarpus*, where the domestic house plant is the result of a series of hybridizations between different species, the florist's gloxinia (*Sinningia speciosa*) was collected from one founder population near Rio de Janeiro in Brazil after 1815. All the subsequent diversity in colour and flower form has been released from that one original wild population over the last two hundred years.[31]

Today, with easy access to incredible plants and a media that saturates us with the exotic, it is difficult to imagine the profound impact of those early introductions. Yet, amazingly, some species have survived this ceaseless denaturing of the exotic. They still fascinate us, and millions of them are grown every year in the nurseries of Florida, Thailand, the Netherlands and California. They are part of an increasingly global industry that has always marched with science to satisfy the demand for novelty.

The first artificial hybrid of *Phalaenopsis*, commemorating the genius plant breeder John Seden of Veitch Nurseries, from James H. Veitch, *Hortus Veitchii* (1906).

two

Monsters and Beauties:
Breeding a Better House Plant

🪷

'Better to breed for what plants have always offered – nonhuman beauty,
originality, slowness, and creative power . . . The art in ornamental plant
breeding resides less in domination of materials, of which kitsch provides
countless examples, than in curiosity, wonder, and love of materials.'

GEORGE GESSERT, 2012[1]

Over thousands of years we have modified plants to provide food and fibre. We are instinctive fiddlers with genetic diversity; without knowing the genetic mechanisms for plant breeding, successive civilizations have created the crops that feed us today. However, there is something different about the breeding of house plants. While it may be driven by a quantifiable utility (such as number or size of flowers), it is also driven by the eye of the breeder and his or her individual interpretation of what is both beautiful and market-worthy. The house-plant industry is built on the fact that people like having plants in their homes, and that they are regularly tempted to buy more of those plants.

The breeding of food plants focuses on improving a dangerously small number of crop species, many of which have an agricultural heritage that stretches back over thousands of years. In contrast, the breeding of new house plants involves experimenting with hundreds of species and many thousands of cultivars. With a few notable exceptions – such as the ancient domesticates of dahlia, paeony and chrysanthemum – many of these plants have a cultivated history of

65

Flaming Katy, *Kalanchoe × blossfeldiana*, a hybrid plant derived from several wild Malagasy species.

less than a century; in some cases, a few decades. In the words of the bio-artist George Gessert, 'the great age of [plant] domestication was not sometime in the distant past, it is now.'[2]

The process of domestication for a house plant takes the original wild collection and modifies the plant's physiology and morphology into a form that can survive the conditions of the house, that can be propagated efficiently and transported with minimum mortality, and that will attract the eye of the purchaser. In this light it can be viewed as both a commercial venture and an ongoing artistic endeavour, the forging of a new set of living performance pieces. Jean des Esseintes, the aristocratic collector of poisonous plants in Huysmans' novel *Against the Grain* (1884), comments that 'in these present times the gardeners are the only and true artists.' In July 1936 the renowned photographer, horticulturist and plant breeder Edward Steichen exhibited his delphinium hybrids at the Museum of Modern Art in New York; this was an art installation, not a garden show.[3] The breeding and selection of new house plants constitute an artistic endeavour as much as a sculpture or painting does, and, as with all artistic endeavours, the beauty is in the eye of the beholder. Horticulture, while the slowest, is arguably among the most sophisticated of the performing arts.

House-plant breeding operates in that fascinating, sometimes troubling, but undeniably fertile zone where art, ethics and commerce overlap. There has been an ethical dilemma associated with it for centuries. The London nurseryman, and author of *The City Gardener* (1722), Thomas Fairchild is rightly recognized as a pioneering nurseryman and the person who created an early intentional ornamental plant hybrid – probably the first.[4] He crossed the popular ornamentals sweet william (*Dianthus barbatus*) with the carnation (*D. caryophyllus*), his resulting hybrid plant being called Fairchild's 'mule'. He presented this plant to the Royal Society in 1720. It is difficult today to understand the controversy attached to the apparently innocent act of crossing two species. Religious doctrine at the time promoted the idea of the constancy of species, namely that each species was a

The ancestor of the domestic African violet, *Saintpaulia ionantha*, from the Usambara Mountains of Tanzania.

fixed entity as forged at the point of Creation.[5] Fairchild, having demonstrated that a new plant could be bred, was worried that he had placed his soul in mortal danger.

Farmers throughout the ages have spotted and salvaged the occasional sport, those spontaneous mutations in a cultivated stock of plants that create a redder apple or fuller grain. But it took some eighteenth-century pioneers to open the door to hybridization as a tool for routinely breeding new plants.[6] Josef Gottlieb Kölreuter (1733–1806), who had a vision that hybrids could be better and bigger plants, started crossing work with *Dianthus* and other hardy garden plants. He was followed by Karl Friedrich von Gärtner (1772–1850), who over 25 years carried out more than 10,000 crossing experiments. In the United Kingdom, Thomas Andrew Knight (1759–1838) and William Herbert (1778–1847) started trial hybridizations. Herbert, as a senior member of the Church of England, was useful in damping down the religious debate on hybridization.

However, even as late as 1899 there was discussion about artificial hybrids. The botanist and taxonomist Maxwell T. Masters wanted to settle any worries about the theological risks of creating hybrids when he addressed the International Conference on Hybridization and Cross Breeding that year, advising people who worried that hybridizing was 'impious interference with the laws of nature' that no less a figure than William Herbert had hybridized daffodils. He described how London nurserymen were creating hybrids of the Cape heaths but selling them labelled as wild species because they feared offending 'over sensitive religious persons'.[7]

Herbert played a pivotally important role in cultivating and studying the amaryllis, *Hippeastrum*, now a valued house plant. He noted that he could not cross the South African belladonna lily, *Amaryllis belladonna*, with the South American amaryllis, the latter of which he subsequently described as belonging to their own genus, *Hippeastrum*.[8] The first *Hippeastrum* hybrid was produced by Mr Arthur Johnson, always described as a watchmaker from Lancashire rather than as a talented plant breeder, who in 1799 crossed *H. reginae* with

H. vittatum to create *H. × johnsonii*.[9] The 220 or so years after Johnson's cross have resulted in a dazzling range of hippeastrum cultivars, with huge variation in flower size and structure, colour and scent. Successive generations of amateurs, scientists and commercial breeders have taken advantage of a number of important opportunities.[10] The wild hippeastrum species are both interfertile and diverse in flower shape and colour, and occupy a wide geographical range (Brazil, Peru, Argentina and Bolivia) that encompasses habitats at different altitudes, so the raw materials for plant breeding could not be better. In addition, the bulbs could be easily transported from South America to Europe. This gave breeders from the nineteenth century onwards access to a range of wild genetic diversity for incorporation into breeding programmes; this, in turn, coincided with the expansion of hothouse horticulture in Europe and allowed both nursery companies such as Veitch and the owners of private estates to breed many new cultivars. The wild species had been introduced early into cultivation (from the 1690s onwards) and had proved popular, so people knew how to grow them. Today hippeastrums are bred throughout the world, including the Netherlands, South Africa, Japan, Brazil and the United States, and the work is undertaken by a mix of university research teams (such as the University of Florida), government institutions (such as the Instituto Agronômico of Campinas, Brazil, and the United States Department of Agriculture) and commercial companies including Miyake of Japan, Penning of the Netherlands and Hadeco of South Africa.

The great nineteenth-century nurseries were first and foremost gathering houses that absorbed the incredible floods of new plants collected from all over the world. This was a fantastic, serendipitous and chaotic cascade of plants that had caught the eye of their agents and collectors. It is easy to imagine the selection process at work once the material arrived at the nursery. Field notes and sketches would have been scrutinized and collectors interviewed, and a plan would start to develop that assessed germination needs, growing conditions, breeding opportunities and potential markets. Seed was handed over – the value of the potential plant influencing the choice of propagator (dull and unpromising stuff would likely go to the apprentice) – and seed trays were watched anxiously for the first sprouts. Already the breeders would be nurturing their ambitions for the plant and imagining a front cover of the company catalogue or a medal from a prestigious flower show.

Some of the plant material imported was the product of the keen eye of generations of growers in the source cultures. For instance, poinsettia was probably improved by the original Aztec harvesters, Belgian azaleas built on a long history of domestication in Asia, and some of the cordylines marketed by Veitch as dracaena were likely selected by Polynesian cultivators.

Once the seed had been germinated and the cuttings rooted, the plants would have been handed over to the growers. Careful eyes would watch the growth, recording the time to flowering and the size and colour of flowers. Some batches would be discarded with muttered questioning of the collector's judgement, while others would be marked for improvement. The promising stock would be moved to locked glasshouses out of sight and away from snooping competitors. Every batch of seedlings and cuttings was scanned for an improved form – a new variant that merited a cultivar name. Every time a batch of seedlings was raised, the genetic dice had been rolled, and a new variation could result. With recurrent propagation from cuttings there was a chance of a random somaclonal variant, a sport – often manifest as a variegated plant – that could be marketed. Every season's catalogue needed novelties to drive profits and prestige.

The breeders would plan the best way to advance the improvement for the new plant. For example, would it benefit from emphasizing one colour, shortening the internodal lengths or improving the scent? Would you focus on crossing selected individuals within the seed batch or within the same species, or would you hybridize with other species or perhaps even cross with other genera? Crossing is not always easy, and the breeder would have to work out the breeding biology: is it self-incompatible, when is the stigma receptive, does successful pollination depend on a particular set of temperatures or humidity?

The Veitch stable of hybridizers worked on material sent in by their network of field collectors. Contacts with private collections or the curators of botanic gardens would allow the exchange of plant material, so keeping a constant flow into the nurseries. This still happens today, with an informal network of amateur and professional plant breeders who exchange stock and techniques. The Veitch breeders had an eye for a great plant and clearly revelled in the stream of exotic new plants flooding into cultivation. John Seden (1840–1921) of Veitch created many new orchid hybrids (his main claim to fame), and bred new hybrids of many hothouse plants that we now value as house plants, including *Caladium*, *Alocasia*, *Hippeastrum*, *Gloxinia*, *Begonia* and *Echeveria*.[11]

Seden would recognize the plant-breeding work of today – the use of a sharp eye, the intuitive understanding of plant biology and the working towards an idealized outcome. New to him would be the advanced science used today, the global nature of the business and the accelerating DNA revolution. He would also revel in those favoured locations where commercial growers, researchers and amateur horticulturists converge in their interests; it is no accident, for example, that southern Florida is the world centre for breeding and developing tropical aroids and bromeliads.

For a relatively small selection of species, there has been massive and sustained investment in scientific breeding over many decades. Research institutions and universities have played a crucial part in developing certain house-plant crops, such as poinsettia and African violet. The U.S. universities have been key in the breeding of new house plants, for instance the University of Hawaii with the flamingo flower, *Anthurium*, starting in the 1950s with the pioneer breeder Haruyuki Kamemoto. (*Anthurium* has had a long history in Hawaii: the three first plants – one pink, one red, one white – were brought to the islands by Samuel Mills Damon II in 1889.) The University of Florida has played a pivotal role in supporting the U.S. house-plant

Sansevieria, originally from sub-Saharan Africa and now grown globally, is cherished as a house plant throughout the world and reviled as an aggressive invasive weed in places such as Mauritius, Hawaii and Florida.

trade through many decades of research into tropical foliage plants such as the aroids and palms.

House-plant breeding involves a variety of approaches. The simplest is the selection of plants from existing cultivated populations. For instance, with *Sansevieria* (mother-in-law's tongue or snake plant), trials for commercial fibre production were established in Florida and Hawaii by the USDA and others from the early 1900s. From these trial plantings keen-eyed horticulturists spotted 'good' plants that were subsequently given cultivar names and introduced into the house-plant market; for example, *S.* 'Koko' and *S.* 'Alva' were selected from trials in Koko Crater near Honolulu. The India rubber plant, *Ficus elastica*, was subject to a colonial-era commercial trade in its rubber-like sap, *caoutchac*, and was a common element in botanic garden collections before its adoption as a house plant in the mid-twentieth century.[12]

For many species, new cultivars have been selected as 'sports', spontaneous mutations spotted in large batches of cuttings. The dragon trees, *Dracaena*, closely related to *Sansevieria*, are a long-established indoor plant. They are tough, tolerant of low light and easily grown from large stem cuttings. The majority of new cultivars are derived from sports. For instance, *D. fragrans*, originally from West Africa, has produced 'Kanzi', 'Jelle', 'Lemon Surprise', 'Golden Coast', 'White Jewel' and 'Janet Craig Gomezii'.[13] The same is true of the figs, for example *Ficus benjamina*, with a wild distribution from India to southern China. The cultivars 'Indigo' and 'Midnight' were sports of *F. benjamina* 'Exotica'.[14] However, some species are very stable and,

Dracaena fragrans, one of the world's most popular house plants, as seen in *L'Illustration horticole*, vol. XXVII (1880).

despite being propagated from cuttings by the million, have yet to yield a new commercial cultivar. Notable among such species is the popular heartleaf philodendron (*Philodendron scandens*).

In *Dieffenbachia*, the dumb cane – a favourite house plant since Victorian days, with the first hybrid made in 1870 – new cultivars are still produced by hybridization and from the generation of novel mutations (somaclonal variation) induced by micropropagation.[15] A number of patented cultivars were selected from spontaneous

The current range of *Dieffenbachia* cultivars has resulted from a process of hybridization and inducing somaclonal variation. From *L'Illustration horticole*, vol. XXX (1883).

The 'Hibotan', or 'Moon cactus', with luminescent orange or yellow cactus globes sitting on short green stalks, is a graft between two species: a sport or mutant with no photosynthetic tissue of *Gymnocalycium*, the root stock a cutting of the climbing cactus *Hylocereus*.

mutations; for example, 'Tropic Snow' is a sport of *D. amoena*, while 'Tropic Sun' and 'Maroba' are sports of 'Tropic Snow'. The University of Florida has also produced a range of interspecific hybrids, including 'Triumph', 'Victory', 'Tropic Star', 'Starry Nights', 'Star White', 'Star Bright', 'Sparkles', 'Tropic Honey' and 'Sterling'.

Plant breeders use hybridization between species in the same genus to obtain a desired characteristic found in a related species, for instance crossing the potted cyclamen, *Cyclamen persicum*, with the wild *C. purpurascens* to produce scented cultivars. However, the creation of yellow-flowered cyclamen cultivars, something that does not occur naturally, depends on the use of new technology such as ion beam irradiation.[16]

Succulents are popular house plants, with their attractive architectural structures and ability to tolerate physiological stress. Among the most readily available are the luminescent orange or yellow cactus globes on short green stalks sold as 'Hibotan' or 'Moon cactus'. These are grafts of two species: the scion (top) is a sport or mutant of *Gymnocalycium* with no photosynthetic tissue; the rootstock (stem) is often a cutting of the climbing cactus *Hylocereus*.[17] It is extraordinary to think that every year around 10 million of these nasty things are sold around the world, most of them produced in the nurseries of South Korea. One succulent group that has seen a recent resurgence in popularity is *Echeveria*, historically valued for its use in municipal carpet bedding and floral clocks, but now favoured as a house plant. There are about 140 species of *Echeveria*, all native to the Americas, from Texas south to Argentina; the highest diversity of species is native to the mountainous areas of southern Mexico. Breeders have crossed species to make new hybrids and have also crossed *Echeveria* with other genera in the same family, the Crassulaceae.[18] Accordingly, we have × *Graptophytum* (a cross between *Graptopetalum* and *Echeveria*), × *Pachyveria* (a cross between *Pachyphytum* and *Echeveria*) and × *Sedeveria* (a cross between *Sedum* and *Echeveria*). To add another layer of complexity, you can also backcross, so × *Sedeveria* can be crossed with *Echeveria*.

Extraordinarily complex hybrids are produced by orchid hybridists. For instance, *Potinara* is a nothogenus (a new genus resulting from artificial hybridization) comprising intergeneric hybrids between four different genera, *Brassavola*, *Cattleya*, *Laelia* and *Sophronitis*.

The original seed batch of African violet (*Saintpaulia ionantha*) collected from Tanzania has formed the basis of an extraordinary and ongoing experiment in domestication. Like that of so many house plants that have been intensively bred, its history is one of investment by individual breeders, teams of scientific researchers, the availability of new breeding techniques and the demands of a market driven by novelty. The much-loved African violet has over the last 120 years been transformed by breeders, morphing the original wild collections into mass-produced plants with thousands of different cultivars, each exhibiting variations in flower structure, flower colour, leaf form and growing pattern. There were two early collections of the African violet that failed to be recognized as a new genus, the first by John Kirk in 1884 and the second in 1887 by Rev. W. E. Taylor. In 1891 the German colonial administrator Baron

Walter von Saint Paul-Illaire (1860–1940) sent seed collected in the Usambara Mountains of Tanzania (then German East Africa) to his father, Hofmarschal Baron Ulrich, who in 1892 provided material to the famous botanist Hermann Wendland of the Herrenhausen Gardens in Hanover. These plants were exhibited under the trade name of Usambara Veilchen (Usambara violets) at the International Horticultural Exhibition at Ghent in 1893. This was the point when the African violet was released to horticulture. The beautiful lithophyte from the limestone hills and remote mountains of Tanzania and southern Kenya was on a trajectory to become perhaps the ultimate house plant, grown on millions of kitchen windowsills and transformed profoundly in shape, character and value. It is easily propagated by both seed and vegetative means (such as leaf cuttings), and soon produces flowers, so breeding quickly produces results and allows the rapid assessment of the plants raised. Importantly, it is a species where both amateurs and professionals have played a key role in developing the modern plant.

The first phase of breeding was based on crossing between plants of *S. ionantha* and the raising of large seed batches from which promising plants could be selected. African violets entered the U.S. trade quickly after their debut in Ghent. The New York florist and bulb dealer George Stumpp imported plants from Germany in 1893 or 1894. He in turn may have supplied Roger Peterson of Philadelphia, an early commercial grower in the first decade of the twentieth century. However, the breeding of new cultivars can in large part be traced to the production of the 'Original 10 Crosses' raised from seed sourced from Germany and the United Kingdom by the Armacost and Royston nursery in Los Angeles, which released ten selected cultivars in 1936. A decade later, in November 1946, the first African Violet Show was held in Atlanta, Georgia, comprising two hundred exhibitors from eleven states, exhibiting 31 cultivars. Since then the cult of the African violet has spread across the world, with breeders and hobbyists in China, Ukraine, Poland, Korea, Japan and Russia all pursuing their ideals of beauty. It is intriguing to think of those first

Perhaps more than any other house plant, the African violet has been transformed by domestication.

ten hybrids, a product of U.S. commerce, conquering the Cold War windowsills of the Eastern Bloc. The most extreme African violets are produced today by amateur breeders in Russia and Ukraine, creating extraordinary plants with large masses of brightly coloured flowers, including greens and yellows, often ruffled with different-coloured edges.

The 'Cold War phase' of African violet breeding in the 1950s used a set of new techniques.[19] With plants raised *in vitro*, that is, grown on sterile agar in sterile glass containers, a variety of chemical treatments were employed to create novelties by random mutation. Natural mutagenic agents, such as colchicine and caffeine, were applied. Chemical mutagenesis (for instance using ethyl methanesulfonate) was also used, usually in combination with colchicine or radiation. The colchicine was used to create polyploidy (the duplication of the original chromosome numbers), often resulting in changes to growth form, such as dwarf plants with smaller leaves that are suited to table-tops. Variegated sports were also produced from tissue-cultured plants.

Much of this early breeding was a numbers game. It was difficult to predict what sort of mutation would occur, and only a small number of the surviving mutations would be stable enough to maintain their new characteristics and subsequently be used as new breeding material. During the 1950s and '60s there was much interest in using the newest nuclear technology to grow new cultivars. The Burpee Seed Company, following pioneering work using colchicine to improve marigolds (the Tetra Marigold, in 1940), took up the challenge of improving the garden annual *Zinnia* with enthusiasm. Seed was bombarded with X-rays, with no results; then the zinnia field was fertilized with radioactive phosphorus, and then the plot sprayed with colchicine. Not surprisingly, new variation resulted – only in the plants, it is to be hoped.[20] 'Atomic Gardens' were staged at U.S. flower shows, exhibiting radioactively improved house plants, flowers and vegetables, including African violets; in addition, 'Atomic Seeds' and 'atomic energized potting soil' were for sale.[21]

House plants provide a focal point for social activities and competitive events, as shown by this U.S. African Violet Show in 1949.

As molecular technology grew, breeders gained a deep under-standing of the African violet's genetics and the mechanisms that influenced flower development and overall morphology. Some of the early work in creating transgenic African violets was driven by the need to establish resistance to root-knot nematodes, using a bacte-rium, *Agrobacterium*, to move a specific gene for resistance into target plants. With the advent of CRISPR (Clustered Regularly Interspaced Short Palindromic Repeats) technology, it is now possible to pre-cisely target or 'edit' specific genes that occur naturally in the plant to generate modified plants without introducing foreign genetic material.

During the 1950s plant breeders used exposure to radiation to promote mutations, and the resulting plants were promoted as 'Atomic Energized'.

The domestication of the orchid as a house plant has converted a very expensive exotic product that was the preserve of the rich into a readily available and affordable purchase that will thrive in a kitchen window. In terms of conversion from elite indulgence to supermarket purchase it bears comparison to the availability of pineapple – once an expensive luxury, now a tinned or refrigerated product, still loved but with a somewhat compromised mystique.

The moth orchid, *Phalaenopsis*, native to tropical Asia, exemplifies the process of domestication that leads to new house plants. James Veitch in his *Manual of Orchidaceous Plants* (1894) stated that 'the introduction of species of *Phalaenopsis* . . . has been one of the most difficult cultural problems horticulturists have been called upon to solve.'[22] They were beautiful plants, however, and the orchid breeders took up the challenge.[23]

Veitch's nursery produced the world's first hybrid orchid in 1853, with the resulting plant, *Calanthe × dominyi*, flowering in 1856; this was followed by a hybrid *Cattleya* flowering in 1859 and a hybrid *Paphiopedilum* flowering in 1869. The first hybrid *Phalaenopsis* was produced by Veitch's in-house wizard, John Seden, who made a cross between *P. amabilis* and *P. equestris*. This was a risky venture as the successful raising of orchids from seed was not guaranteed. Following the cross in 1875 a single resultant seedling was raised to flower in 1886. This opened the door and Seden was quickly producing new hybrids. By 1900 he had raised an additional thirteen hybrids to flowering. The ability to cross species was established and the hybridizers were ready to explore the commercial potential of their artistry. However, the raising of orchids from seed to flowering was difficult and could not be adopted on a commercial scale. The orchid seed was sown onto fern fibre and other natural substrates, and germination was both slow and erratic.

The relationship between the orchid seed and fungi was cracked simultaneously in 1909 by the French botanist Noël Bernard and the German Hans Burgeff, who discovered that orchid seed germinates in association with fungi that live in the soil. The discovery of this symbiotic relationship was applied commercially by the collector and nurseryman Joseph Charlesworth (*c.* 1851–1920) and the mycologist John Ramsbottom (1885–1974); together they created a ground-breaking system of *in vitro* propagation (sterilized peat inoculated with the fungus in glass bottles), improving the survival rate for raising seedlings. In reality, this was an early stage in the evolution of biotechnology, the early commercial production of plants in sterile laboratory conditions.

In the 1920s Lewis Knudson at Cornell University worked out that the symbiotic fungus could be replaced with a mix of agar jelly and the salts and sugars that would feed the orchid seedlings (the famous Knudson C medium). Commercial production of seedlings jumped forwards again as this new method dramatically improved the survival of seedlings. While this breakthrough allowed the breeding of many thousands of hybrids, there was a problem in replicating individual plants, particularly those highly valued selected clones that had been selected from the seed batches. In 1949 Knudson's colleague Gavino Rotor successfully grew young clonal plants of *Phalaenopsis* through tissue culture, using small pieces of the orchid flower nodes

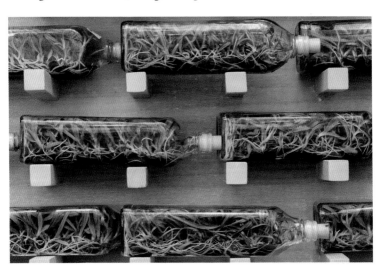

The development of in-vitro (laboratory) propagation for commercial orchid propagation has transformed a luxury into a supermarket product.

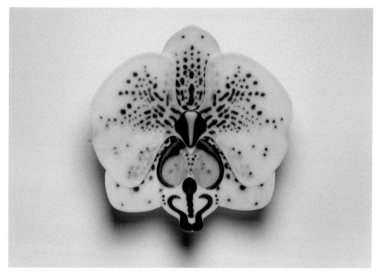

Laura Hart, *Phalaenopsis Violet Appaloosa*, 2018, glass.

to propagate hundreds of clonal duplicates. Today millions of plants are produced by this process, known as micropropagation.

Hybridists were now breeding and raising new *Phalaenopsis* hybrids better suited to the house, but there was a need to develop the best commercial growing conditions and to understand the physiology of the crop. This was an international initiative, and universities from all over the world undertook research from the 1960s onwards. Important players included the USDA, the University of Florida, Cornell University and Texas A&M University in the United States; Osaka, Miyazaki, Nagoya and Nihon universities in Japan; Chiba University in South Korea; and the Hebrew University in Israel. As a house plant, *Phalaenopsis* is a huge success, offering a variety of differently sized plants (including miniatures), a range of colours and, above all, the ability to thrive and flower in an ordinary home. The production of *Phalaenopsis* is now global. A new hybrid raised by a U.S. breeder may involve many partners during its production: Japan for micropropagation, cultivation in China for bulking up the stock, growing of the *in vitro* plantlets in the Netherlands, and their final return to the United States for sale as flowering plants. This beautiful group of plants has come a long way since that first hybrid by John Seden of Veitch.

For a long time the breeding of house plants had been a process of removing plant material from its country of origin and establishing ownership by the overseas breeder. Accordingly, while a *Philodendron* or *Saintpaulia* cultivar may be derived from a wild collection in Brazil or Tanzania respectively, the new cultivar may be owned by a U.S. or Dutch company, through a set of legal processes that protect the commercial interests of the breeder. For instance, in the United States many new house plants are patented through the Plant Patent Act of 1930. Here is the dilemma: we fill our houses with plants ultimately derived from the tropics, in some cases legacies of ancient cultures, and we put commercial value on the investments made by European or American breeders. This business model does not take into account cultural legacies or allow for the conservation of habitats that hold wild populations of commercial species; for example, Tanzania gets no funding from the cultivation and sale of African violets in Europe and North America.

The wave of investment in modern house-plant breeding started in nineteenth-century Europe, rolled over to the United States (California and Florida) and is now firmly established in Asia (Thailand, China, Korea and Japan). A tour of any plant market in Thailand will provide an intoxicating indication of the scale and scope of new breeding, clearly showing that each new set of breeders enriches the process of domestication through innovation and creative imagination. We are seeing Mexico reclaim the poinsettia, and we can hope that the next wave will be in nations such as Colombia or South Africa, where breeders are working with their own horticultural and botanical assets to create house plants for national and regional consumers.

Not just a spider plant, but a small ecosystem of symbiotic life forms.

three
Health, Happiness and Mutualism

☸

'Plants seem absent, as though lost in a long, deaf, chemical dream. They don't have senses, but they are far from being shut in on themselves: no other being adheres to the world that surrounds it more than plants do . . . they participate in the world in its totality in everything they meet.'
EMANUELE COCCIA, 2019[1]

House plants may seem to be passive additions to the home. In fact, they are living things that interact with us and the ecology of our living spaces. More than half of the world's population now live in cities, and it is predicted that nearly 70 per cent of people will be living in urban environments by 2050.[2] This is a big and expanding habitat; for instance, while Manhattan may have a surface area of 59 square kilometres (17 sq. mi.), its indoor biome (a measure of floor space) covers 172 square kilometres (50 sq. mi.).[3] This is the space where we live (since we spend about 80 per cent of our time indoors), where we work and relax, and where we grow our house plants. Despite what we may imagine, our homes are an evolving habitat with new ecological processes, both for us and for an increasing range of other cohabiting species. Some of these species we actively select and cherish, and others are there as uninvited colonists.

The mutualism between people and house plants is entering a complex and potentially more intimate phase. Our homes are not sterile, sealed units, and there is a flow of life between the indoor

87

volumes of our home and the outside world. Some of this flow is intentional – we deliberately introduce life including pets, plants, living fermented products and indeed other humans into our homes. In addition, there is a flow of colonizers coming in through open windows, air-handling systems and water supplies and as hitchhikers on plants, pets and ourselves.

Thousands, probably hundreds of thousands, of species share our homes. A study of North Carolina homes found representatives of life's three great domains, the eukaryotes (which includes insects, plants, fungi and so on), more than 8,000 bacteria and a surprising diversity of archaea taxa.[4] Most of this research has focused

on examining the life flourishing in extreme environments such as washing machines and hot-water systems, and the potentially 'fertile' habitats such as kitchens and toilets, but the role of house plants in the indoor biome is surprisingly poorly studied. A cost of spending more time indoors is that we are no longer exposed to the full spectrum of microbial diversity found in natural outdoor environments. This profound change is thought to have disrupted our ability to acquire a functional adult microbiome and resulted in a range of autoimmune disorders and allergies.

You could argue that each house plant, a living plant growing in a soil mix, represents a biodiversity hotspot in the indoor biome, a

Concert for the Biocene, an installation by Eugenio Ampudio at Gran Teatre del Liceu, Barcelona, 2020.

tiny Galápagos in a Pacific Ocean of carpets, tiles and plaster. Each house plant has a living film of fungi and bacteria on the leaves (the phyllosphere), a set of viruses living in the plant tissue, a set of fungi, archaea and bacteria in the soil or compost (the rhizosphere) and the macroscopic biota, the aphids, red spider mites or mealy-bugs. A house plant from even the most modern and apparently hygienic industrial nursery is a complex living system, not just a plant. Like a human, a house plant is a meta-organism, the plant's physical structure and genome wrapped up with a wide variety of intimate co-inhabitants.[5]

Different house-plant species support different sets of bacterial diversity. Researchers suggest that these mini ecosystems have the potential to influence the overall indoor microbiome by increasing its diversity and filtering airborne microbes. A study of the microbiome of the cosmopolitan spider plant, *Chlorophytum comosum*, indicates that the plant's microbiome spreads into the home environment and probably interacts with both the human and other plant microbiomes.[6] Potentially, we can use house plants as a source of microbial biodiversity and possibly beneficial micro-organisms, with plant-associated bacteria providing a defence against pathogens by stabilizing indoor microbial ecosystems, enhancing overall biodiversity and reducing outbreaks of pathogens.[7] For instance, a high proportion of phyllosphere-associated bacteria can produce volatile organic compounds (VOCs), many of which are active against the fungal plant pathogen *Botrytis cinerea*.[8]

Increasing urbanization brings its own challenges in terms of human health. Non-communicable diseases such as diabetes, cardio-vascular disease, cancer and depression are now among the fastest-growing health challenges around the world. While many factors are involved in this increase, lifestyle factors such as physical activity, diet and stress are particularly important.[9] There is increasing evidence that lack of access to nature in cities, and associated sedentary indoor lifestyles, is linked with physical and mental health disorders including vitamin D deficiency, asthma, anxiety and depression.[10]

Each house plant is an island of life that interacts with the ecology of the home.

Better access to nature has been shown to improve both physical and mental health, and recent research indicates that indoor plants bring a specific benefit.[11]

The benefits of house plants towards our well-being are frequently discussed, and there appears to be an intuitive understanding that plants are a 'good thing'. The Victorians, however, had ambivalent views on plants in houses. There was an idea that plants could give off a dangerous and toxic 'effluvium'.[12] Jane Loudon wrote in the *Ladies Magazine of Gardening* in 1841 that plants in the dark 'give out carbonic acid . . . and as a superabundance of this gas produces stupor, head-aches, and a sense of suffocation in those that breathe it, plants often produce these evil effects on those who keep them in bedrooms'.[13] This is despite the great Enlightenment scientist Joseph Priestley stating in an address to the Royal Society in 1772 'that plants, instead of affecting the air in the same manner with animal respiration, reverse the effects of breathing, and tend to keep the atmosphere sweet and wholesome, which it had become noxious, in consequence of animals living and breathing, or dying and putrefying in it'.[14]

A vast number of websites, YouTube videos and books inform us, with varying degrees of scientific credibility, that house plants and cut flowers can improve our well-being by improving air quality, strengthening our ability to concentrate and to complete creative tasks, and combating the impact of stress and depression. Studies have repeatedly shown positive effects from exposure to plants and nature; for instance, people recover faster from surgery in the presence of vegetation, have increased cognitive functioning after forty-second views of vegetation, and have increased self-esteem from participation in 'green exercise' programmes.[15] Several comprehensive reviews of the health benefits of nature have recently been published, all illustrating the varied benefits of plants and vegetation.[16]

House plants are actively promoted as mechanisms to improve indoor air quality by removing damaging VOCs from the air, pollutants that may be released directly from building materials and furnishing, and from other sources including air fresheners and cooking. These are a primary contributor to the symptoms of 'sick building syndrome' and other health problems associated with poor-quality indoor air. In the late 1960s the environmental scientist Bill Wolverton led a team that discovered that aquatic plants filtered the herbicide Agent Orange out of habitats. As a result, the National Aeronautics and Space Administration (NASA) gave Wolverton funding to explore the use of plants' roots to filter contaminants from air supplies during deep space exploration. The result, based on 27 years of research, was 'The NASA Guide to Air-Filtering Houseplants', a list of plants identified for their ability to remove carbon-based contaminants and VOCs from the air. As well as non-toxicity to human plant owners, other criteria included ease of growth and extended lifespan.[17] This document has spawned thousands of websites, YouTube videos and marketing campaigns all promoting indoor plants as tools for controlling indoor air pollution.

Research has demonstrated that house plants can reduce the concentration of VOCs in the atmosphere, and that they do it through three main processes: (a) removal by the above-ground parts of the plants; (b) removal by the roots; and (c) removal by the micro-organisms and organic matter in the compost. The key message is

Sansevieria trifasciata, among the toughest of all house plants and mythologized as a pollution eater.

that slowly decaying house plants do not absorb pollutants. For the efficient removal of pollutants you need actively growing plants with strong root growth, and huge volumes of them.

The challenge is to translate these experimental studies into a clear understanding of what works for the home or office. Buildings are complicated structures with differing room sizes and fluctuating airflow and temperatures, and the release of VOCs can vary within a single building from day to day. Many of the studies used relatively small, atmospherically sealed research chambers that cannot

Samuel John Peploe, *Aspidistra, c.* 1927, oil on canvas.

be compared with the complexity of a house, and involved only one plant species and one type of pollutant. Such experiments are difficult to translate and provide poor guidance to the effective use of plants to manage indoor pollution.

However, the fundamental message is correct. House plants are good for you, and if planted in big enough volumes and maintained in active growth they may influence the concentration of VOCs in the home. Opening the windows will have more impact than even a massed and thriving planting of *Dracaena* or *Spathiphyllum*. However, more research is needed to improve the efficiency of house plants as instruments for managing indoor air pollution.[18] We know that some plant species are better than others, so opportunities exist to breed and select new types with modified leaf cuticles, stomatal behaviour that changes the flow of air in and out of the leaf, and rates of photosynthesis. Genetic modification can be used to improve efficiency at absorbing pollutants; for instance, a common house plant, the pothos vine (*Epipremnum aureum*), modified with the mammalian cytochrome P450 2E1 gene shows a marked improvement over normal plants in detoxifying two VOCs, benzene and chloroform, suggesting that biofilters using transgenic plants could remove VOCs from home air at useful rates.[19] Above all, to make a meaningful impact on offices and homes we will need to grow far greater volumes of actively growing plants, probably with managed airflow through green walls; sadly, the decaying and chlorotic dracaena, mulched with a teabag, will not suffice.

While we interact on physiological terms with plants, we also interact with them in a way that influences our mental health and sense of well-being. It is clear that we have emotional relationships with house plants.[20] Mary McCarthy's *Birds of America* (1965) explores the relationship between the unfortunate Fats, a × *Fatshedera* plant, and his carer Peter Levi; the plant, 'long, leggy and despondent', is taken on walks through Paris but subsequently dumped. In George Orwell's case the unfortunate aspidistra comes to symbolize his anger at suburbia and 'middle England', a metaphor also used by

H. E. Bates in *An Aspidistra in Babylon* (1960).[21] Flannery O'Connor's short story 'The Geranium' (1946) starts with a geranium (specifically a *Pelargonium*), smashed and abandoned on a New York pavement; the expected rescue by the main character, Old Dudley, does not happen. House plants can provide solace and inspiration, and the rescue of a plant can have a profound impact on a person, producing a sense of pride and achievement. In Haruki Murakami's novel *1Q84* (2009–10) the assassin Aomame, destined for prison, worries about the fate of her potted rubber plant, her only living companion, 'her first experience of living with a thing that had a life of its own'.

As an increasingly urban species we live in environments that are crowded, our opportunities for social interaction and self-identity restricted, and we find it increasingly difficult to connect with nature. Here the house plant plays a key role on several fronts, through the benefits of making a collection (our own cabinet of curiosities) and, importantly, through the benefits of interacting with nature and with like-minded people.

Some households are content with a single house plant, while others fall into the sweet trap of collecting and filling their house with plants. You could dismiss the collecting of plants for decorating a house as the materialistic and luxury consumption of a short-lived asset. You could also argue that it is a benign compulsion that creates delight. The act of purchase transforms the mass-produced product, the house plant, into something valued and possibly loved. Importantly, each additional plant is another step towards completion of the collection (as long as we recognize that no plant collection can ever be complete) and towards defining and reinforcing self-identity.

Much has been written about the psychology of collecting, some of it based on some fairly unsavoury Freudian interpretation. However, visit a plant sale or society meeting and it is very clear that the purchase, collecting and cultivation of house plants brings great pleasure to millions of people. The motives can be diverse: for some it is close to an obsession, for others a passion for a particular plant

Paul Cézanne, *Terracotta Pots and Flowers*, 1891–2, oil on canvas.

group, or the excitement of the hunt, or about making friends, gaining knowledge and building self-esteem and identity.

Given the increasing isolation and conformity of urban life, it is likely that collecting and cultivation (or curation, since these plants are receptacles of memories and stories) provide several services. They provide autonomy – the collection is yours and you curate it as you wish; you build experience, competence and knowledge, which in turn provide a basis for social networking; and you cultivate an eye for diversity and change. For Generation Rent, house plants are affordable pieces of nature that can move from one rental property to the next.

The psychological benefits of gardening are well known, but this has focused on the outdoor environment, and it is useful to explore the role of house plants in promoting mental health. In many parts of the world, as exemplified by the Asian megacities, a large proportion of the population does not have access to a garden, so the quality of the home environment is vitally important.

House plants can be almost family. A single plant can provide personal continuity, a link to the past, the present and perhaps the

House plants continue to define our living space: photograph of *Philodendron scandens*, 1959.

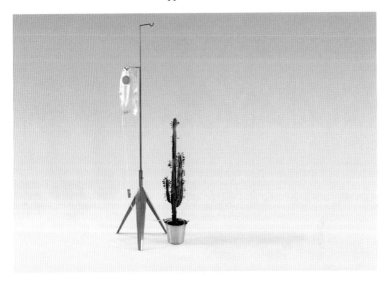

House-plant watering system designed for busy people, by Keita Augstkalne, 2018.

future. Often house plants carry a personal history: a gift or a trade with a community of friends, a legacy passed between generations or a memory of another home or time. They may be looked after for decades, their offspring shared with family and friends, and indeed may pass from one generation to the next. They can move with migrants, fragments of green that connect the dispossessed with home. Mint plants have moved with Middle Eastern families across Europe, providing a memory of home, each cultivar an expression of a particular origin and food tradition. In Miami, many Cuban families have a pot of Cuban oregano, *Plectranthus amboinicus*, on the windowsill or veranda, a plant that speaks of home and identity. Yet the Cuban oregano is native to east Africa, growing wild from Kenya south to KwaZulu-Natal, from where it spread across the Indian Ocean trade routes and eventually to the Caribbean.[22] Each successive community has adopted it as part of their identity, and it has provided solace to each diaspora.

House plants and their paraphernalia are gardens and habitats in microcosm, and carry crystallized echoes of bigger habitats and distant climes. While they may not require the strenuous involvement

99

Lucian Freud, *Interior with Plant, Reflection Listening (Self-Portrait)*, 1967–8, oil on canvas.

of an outdoor garden or allotment, they provide opportunities for creativity and allow people to personalize their homes. This is of great value to an urban population. The choice of plant, of container or terrarium, and of the way a home is decorated makes people feel 'at home' in their space. You may occupy a garden for a few hours each week, but you sleep, eat and relax in your home, and it is a manifestation of your identity and personality. Personalizing a home with plants can make it idiosyncratic and distinctive, strengthening a sense of individuality and boosting self-esteem. This sense of social identity has been promoted by active online communities for house-plant enthusiasts. That is not to say that all relationships with house plants are positive, however; these companions in our homes can be the focus of damaging attentions, as Orwell wrote:

Gordon had a sort of secret feud with the aspidistra. Many a time he had furtively tried to kill it — starving it of water, grinding hot cigarette-ends against its stem, even mixing salt with its earth … After fuelling his stove, he even deliberately wiped his kerosiny fingers on the aspidistra leaves.[23]

Pottering, while often derided as a somewhat Wodehousian escape mechanism, is an invaluable and positive psychological process. It is a key part of all plant care — the process of inspection, a ceremony of fiddling and picking over, that can take a few minutes or can absorb hours. This time of calm focus, when the attention shrinks to a smaller-scale world, is referred to as an attention-holding activity.[24] The all-encompassing focus of looking after plants, contemplating their needs and their growth and anticipating future growth and flowering is a period when time slows down and you recharge your mental batteries.

As our lives become more complicated and technology dominates our cultural, business and personal interaction, it is strange that a small pot of a photosynthesizing organism should provide so much solace. This is attributed to our innate need for nature, what E. O. Wilson called 'biophilia', and supposes that we are attracted to other life forms, we seek them out. Once this was based on survival, the plants, animals and landscapes that provided our living space.[25] Now we take comfort in and derive inspiration from other living organisms. The counter-theory, that of biophobia, recognizes a fear of certain components of biodiversity, species that are perceived as distasteful or a threat, such as, in this context, scale insects and mealybugs.

Evolving from Wilson's theory of biophilia is the idea of biophilic design, an approach to improve our well-being by promoting connections with nature in offices, homes and public spaces. It uses shapes and forms that are found in nature, natural ventilation and building materials, plants, extensive natural lighting, views to the outdoors and a merging of the internal and external landscape.[26] Research has shown that plants in the workspace can significantly improve the morale of

staff and promote well-being and performance, including reducing stress.[27] A study of Australian office dynamics revealed how plants (the 'desk buddies') were given names and the behaviour of the plant became the focus of office conversation; the study makes no reference to any stress or sense of shame associated with a dead or moribund plant.[28]

This mutualism between ornamental plant and humanity is creating a new set of environments at a variety of scales.[29] In the home, plants will continue to be valued as a personal enrichment, a set of specimens to enjoy and value. Those same plants will increasingly form part of the urban fabric as indoor spaces become more biophilic, more alive. This is a grand experiment; the urban green walls have the potential to be long-term habitats that mature and diversify over decades, patches of increasing ecological spontaneity. Meanwhile,

Exterior (*opposite*) and interior (*above*) of the Stepping Park House, Ho Chi Minh City, designed by Vo Trong Nghia Architects.

house plants will continue to be valued inside the house — they may be manipulated as agents of a managed microbiome, or engineered to capture more pollutants, but fundamentally they will be loved because they delight, and they look wonderful.

four
The Crystal Legacy of Dr Ward
☘

'To have real verdure in the freshness of its original strength and life,
there is but one method, and that is by the culture of it in Wardian Cases.'
SHIRLEY HIBBERD, 1856[1]

We love putting biodiversity into glass boxes; in our terrariums, museum dioramas and aquaria we create microcosms of imagined tropical worlds. We can trace this custom back to the ancient cabinets of curiosity, but particularly to some pioneering Victorians. Most notable among them is Dr Nathaniel Bagshaw Ward (1791–1868) of Whitechapel, east London, who by cultivating plants in glass boxes not only built a safe container for transporting them across oceans but created the crystal frame for the tens of thousands of little planted worlds that decorated Victorian parlours and provided wholesome entertainment and education in natural history.

This fascination with little worlds continues in elaborated forms today. To many people of a certain generation there was once a terrarium in the house, usually a wonderful old glass carboy, the glass tinged green, whose interior contained a mesh of *Selaginella*, moss and perhaps a struggling *Fittonia*. Interest in the terrarium is growing again, and the traditions of the Wardian Case and aquarium have merged into the work of aquascapers such as the late Takashi Amano and contemporary artists such as Saša Spačal, Makoto Azuma and Mark Dion. At the same time designers such as Giuseppe Licari and

The terrarium, a tiny glass-encased world.

Technology allows small growing environments
to be established anywhere in the house.

Patrick Blanc are magnifying those miniature worlds into large green rooms and vertical gardens where we are enveloped by indoor plants.

Victorian Britain was intoxicated with glass, the material that created new industrial wonders such as Joseph Paxton's Crystal Palace or Richard Turner's Palm House at Kew. The increased availability of the material after the repeal of the glass tax in 1845 and the desire of botanists and amateur naturalists to grow exotic plants created fertile ground for invention. The leading horticulturist and publisher of the popular *Gardener's Magazine* John Claudius Loudon promoted the conservatory not just as an appendage to the nineteenth-century home, but as a room, part of the house: 'A Greenhouse, Orangery, or Conservatory ought, if possible, to be attached to every suburban residence.'[2]

As with most revolutionary inventions, there were earlier forgotten innovations. Several glass cases had been trialled for the cultivation and transport of plants during the eighteenth and nineteenth

Lovis Corinth, *Woman Reading Near Goldfish Tank*, 1911, oil on canvas.

Illustrations from Nathaniel B. Ward, *On the Growth of Plants in Closely Glazed Cases* (1852).

centuries.[3] For instance, Nikolaus Joseph von Jacquin developed his own cabinets for successfully transporting plants during his expedition (1754–9) to the Caribbean islands on behalf of the Imperial Court in Vienna. In the United Kingdom Professor Allan Alexander Maconochie, of the University of Glasgow, was successfully growing exotic ferns and other plants in a converted fish tank fourteen years before Ward's invention.[4]

It took the complicated character of Dr Ward to transform a botanical research tool into a fashionable addition to the Victorian parlour, and with it to revolutionize colonial agriculture. Ward was a lifelong amateur botanist, with a personal herbarium that contained 25,000 specimens. As a working doctor in a heavily polluted and deeply impoverished part of London's East End, he could clearly see the human cost of industrial pollution. Shirley Hibberd, writing in 1859 (some thirty years after Ward's observations), described the difficulty of gardening in London: 'Thousands of beautiful plants are every spring and summer brought from nurseries round London and sold in the City to undergo the slow death of suffocation . . . from an absorption of soot in the place of air.'[5]

As a boy Ward had travelled to Jamaica, and it is easy to imagine how those memories of tropical abundance haunted him as he tried to grow ferns in the polluted London air. However, a simple natural-history observation inspired him to cultivate plants safely in a protected environment. In the summer of 1829, he wrote,

[I] had buried the chrysalis of a sphinx [hawk moth] in some moist mould contained in a wide-mouthed glass bottle, covered with a lid. In watching the bottle from day to day, I observed that the moisture which, during the heat of the day arose from the mould, condensed on the surface of the glass, and returned whence it came; thus keeping the earth always in the same degree of humidity. About a week prior to the final change of the insect, a seedling fern and a grass made their appearance on the surface of the mould.

I placed the bottle outside the window of my study, a room with a northern aspect, and to my great delight the plants continued to thrive . . . They required no attention of any kind, and there they remained for nearly four years, the grass once flowering, and the fern producing three or four fronds annually.[6]

Wardian Cases were simultaneously roped down on the decks of sailing ships enduring salt and storm, and decorating Victorian parlours and providing the focus of polite and morally informative conversation. Ward worked with the Loddiges nursery to test the application of his case to international plant transportation.[7] In 1833

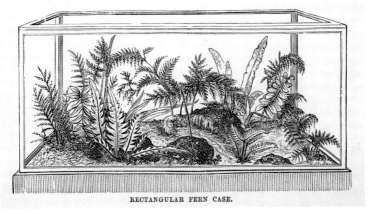

RECTANGULAR FERN CASE.

Illustration of a rectangular fern case, from Shirley Hibberd, *The Fern Garden*, 9th edn (1881).

THE DRAWING-ROOM, GEM, AND ALBERT PLANT CASES.

These have been awarded Prizes by the International Horticultural and Botanical Congress, Royal Horticultural Society, Royal Botanic Society, Crystal Palace, &c.

No. 11.   No. 3.   No. 1.

No. 5.   No. 9.   No. 7.

The above Illustrations represent a few of the numerous elegant Plant Cases and Jardinets we have always on exhibition at our warehouse. (*See also pp. 30 and 31.*)

No. 1.—Tripod Gem Plant Case, furnished with ferns and foliage plants, 63/.; or filled with spring-flowering bulbs, 42/. to 50/.; Table, 42/.

No. 3.—Albert Plant Case, furnished, 18 inches, 65/.; 20 inches, 84/.; 24 inches, 105/.; 30 inches, 147/.; 36 inches, 189/.; empty, 45/., 50/., 70/., 90/., 115/.; Tables 35/. to 50/.

No. 5.—Gem Plant Case, with Rustic Wood Stand Complete, furnished with ferns, &c., 105/.; or filled with spring-flowering bulbs, 84/. to 90/.; empty, 84/.

No. 7.—Rustic Terra Cotta Plant Case, furnished with ferns, &c., 30/. to 35/.; or filled with spring-flowering bulbs, 21/. to 30/.; Table, 42/.

No. 9.—Rustic Terra Cotta Pine Pattern Plant Case, with furnished ferns, &c., 21/., 25/., and 30/.; or filled with spring-flowering bulbs, 15/., 21/., and 25/.; Table, 63/.

No. 9.—Rustic Terra Cotta Toothed Pattern Plant Case on Pedestal, with furnished ferns, &c., 30/. and 35/.; filled with spring-flowering bulbs, 21/. to 30/.

No. 9.—Terra Cotta Plant Case, with furnished ferns, &c., 25/., 30/., 35/., and 42/.; filled with spring-flowering bulbs, 21/. to 30/.

The Cases, when filled with plants, should, if possible, be conveyed under personal superintendence; if required to forward them by rail, we exercise every caution in packing and delivering them safely at any of the London Railway Stations, to be forwarded by passenger train at the Consignee's risk and expense. If filled with bulbs they will carry any distance safely, and are sent carriage paid, if 21/. or upwards in value. As they require to have extra packing, a small charge is made for the case.

LONDON: Printed by Truscott, Son, & Simmons, Suffolk Lane, E.C.

A range of ornate Victorian terrariums, from Barr & Sugden's spring seed catalogue, 1866.

Wardian Cases filled with plants were sent to Sydney. The cases arrived with the plants thriving, despite the ferocious range of temperatures encountered on the voyage. In 1834 the cases were refilled with Australian plants and returned to the United Kingdom. As with the outward voyage, the cases were variously baked and chilled, even subjected to snow, but despite this both George Loddiges and Ward were delighted with the quality of the plants. This proved the case's commercial value, and it was rapidly adopted for widespread use. Three nineteenth-century consignments demonstrate the extraordinary utility of the Wardian Case to the Victorian colonial engine: the plant hunter Robert Fortune successfully shipped tea seedlings from China to India; Brazilian rubber (*Hevea*) seedlings were shipped from Kew to Malaysia to found the Asian rubber industry; and quinine plants (*Cinchona*), a treatment for malaria, were shipped from Peru to the United Kingdom. The Wardian Case provided an international shuttle service between distant agricultural stations and botanic gardens; between 1871 and 1880, an average of 39 were despatched from Kew every year.[8] Wardian Cases also provided an emotional link to the 'old country'. In 1865, when a consignment of primroses was delivered to Melbourne, some 3,000 nostalgic people turned out to see the plants.[9]

Ward became a scientific celebrity, displaying his case at the Linnean Society in 1833 and the Great Exhibition in 1851.[10] The Royal Society, the Society of Arts and the British Association for the Advancement of Science all applauded the innovation.

While Ward was motivated by a horticulturist's instinct to grow plants where they had previously failed, he was also influenced by the prevailing Victorian values of social improvement and religion. He worked surrounded by misery and saw the Wardian Case as an asset to tackle poverty and poor health. A chapter of his book is titled 'On the Application of the "Closed" Plan in Improving the Condition of the Poor', in which he proposes the 'relief of the moral and physical wants of densely crowded populations' by using Wardian Cases to grow nourishing salad greens in window boxes and proposes that the poor could develop businesses supplying plants and models of ruins and towers to decorate the cases. He also considers the value of glass cases and access to sunlight in the treatment of tuberculosis and measles. Ward, as a good Victorian, saw a spiritual value in the Wardian Case and an appreciation of plants. After rhapsodic text on the tree ferns of the Andes and the montane forests of the Canary Islands, he writes:

> the arborescent ferns are the most glorious objects in the vegetable kingdom; and in temperate climes that man is little to be envied who cannot take delight in the phoenix-like beauty thrown over dead and decaying works of Nature and Art, nor be led by these visible things of creation to adore the invisible wisdom and admirable workmanship of Almighty God.[11]

The workings of the Wardian Case mystified some people.[12] Contrary to many people's understanding, the case was not a miraculous fully sealed glass box, and it still needed a measure of horticultural intervention. This confusion caused commentators to criticize the invention; for instance, Hibberd described the 'Wardian theory' as a 'delusion and a snare'. It also stimulated some distinctly metaphysical debate about whether these magical glazed cases could stall the passage of time and slow ageing or halt the decay of cut flowers.[13]

It is easy to establish a link between the Wardian Case and that other, albeit even stiller Victorian tableau, the case of taxidermy. Glass cases full of stuffed birds and animals were characteristic of the Victorian parlour – a dead exotic microcosm. It was not unknown for plant collectors shipping orchids and bromeliads to the United Kingdom to trade in exotica such as hummingbird skins. One famous collector was Ward's friend the nurseryman George Loddiges, whose collection of hummingbirds exceeds two hundred species and is now held by the Natural History Museum in London.[14] In 1835 one of

DICK RADCLYFFE & CO., F.R.H.S.

Seed Merchants and Horticultural Decorators.

128 & 129 *HIGH HOLBORN.*

*Queen Anne Window Cases, Early English Conservatories, Old Style
Aquaria and Fern Cases, Conservatories and Window Cases
in ye Old Style. Registered Designs.*

Seeds.		*Bulbs.*
Horticultural Requisites.		*Garden Requisites.*
Window Cases.		*Window Boxes.*
Ferneries Built & Furnished.		*Conservatories Built & Furnished.*
Plants.		*Ferns.*

*The Fernery in H.R.H. the Prince of Wales's Pavilion at the Paris Exhibition was
executed by* DICK RADCLYFFE & CO.

Horticultural Builders.   Garden Furnishers.

DICK RADCLYFFE & CO., F.R.H.S.

Seed Merchants and Horticultural Decorators.

ILLUSTRATED CATALOGUES GRATIS AND POST FREE.

Advertisement for Dick Radclyffe & Co. in *Ferns and Ferneries* (1880).

Loddiges' collectors, Andrew Matthews, discovered a new genus of hummingbird that was named after his employer. *Loddigesia* is a monotypic genus, that is to say, one that contains only a single species, the marvellous spatuletail, *L. mirabilis*, which is now sadly threatened with extinction. Another great nurseryman and collector of exotica was Sir Harry J. Veitch (1840–1924), who had a fine eye for tropical seashells and ethnographic curiosities; parts of his collection are now in the British Museum in London and the Royal Albert Museum in Exeter.

The invention of the Wardian Case coincided with the extraordinary fashion for collecting and growing native ferns, a trend known as Pteridomania. Victorian Britain was growing in wealth, and for the well-off there was both money and time to engage in hobbies, particularly with a brand-new railway network opening up access to the countryside. The Wardian Case became the receptacle for the craze and placed the hobby right at the heart of middle-class parlours. Ferns were exhibited in a dizzying array of cases to suit all tastes and budgets. Ward had one constructed called 'Tintern Abbey' that measured 2.4 metres (nearly 8 ft) square; it contained a model of the abbey with fifty fern species plus other plants including palms and a camellia.

Charles Kingsley, author of *The Water-Babies* (1863), commented on the craze in his book *Glaucus; or, The Wonders of the Shore* (1854–5):

> Your daughters, perhaps, have been seized with the prevailing 'Pteridomania', and are collecting and buying ferns, with Ward's cases wherein to keep them (for which you have to pay), and wrangling over unpronounceable names of species (which seem to be different in each new Fern-book that they buy), till the Pteridomania seems to you somewhat of a bore.[15]

Ferns were regarded as a subtle and superior object for collecting, as is outlined by William Scott in *The Florist's Manual* of 1899: 'It is not hazardous to say that it is superior minds that have a taste or make a hobby of ferns . . . superior minds far above the common herd.'[16]

Similarly, Hibberd waxes lyrical in *The Fern Garden* (1869) about the value of the indoor fernery:

> where gardens are unknown, and even the graveyards are desecrated by accumulations of filth, the fern case is a boon of priceless value. It is a bit of woodland sealed down with the life of the wood in it, and when unsealed for a moment it gives forth an odour that might delude us into the belief that we had suddenly wafted to some bosky dell where the nodding violet grows.[17]

In the United Kingdom commercial fern collectors scoured the countryside for rare specimens to sell to the metropolitan fern addicts. Rare and localized species became heavily reduced in numbers, including the prized Killarney fern and alpine woodsia, and stories abound of collectors scraping out localities and sending consignments by the tonne up to London markets. High prices were paid for leaf variations and hybrids and, as with all plant crazes, there was a confusion of new cultivar names.

In her botanical guidebook and memoir *Hardy Ferns* (1865), Nona Bellairs wryly called for laws to protect ferns from over-collection: 'The poor ferns, like the wolves of olden times, have a price set upon their heads, and they in like manner will soon disappear. We must have "Fern laws" and preserve them like game.'[18] This moving call for conservation comes directly after her lengthy descriptions of searching the Devon countryside for rare ferns.

Pteridomania soon influenced the public taste in pottery, with Wedgwood, Minton, Royal Worcester, Ridgway, George Jones and others introducing pieces decorated with fern motifs. The Coalbrookdale Company from Shropshire produced magnificent cast-iron furniture, including the popular 'Fern and Blackberry' and 'Osmunda Regalis [royal fern]' garden chairs.

The great Victorian fern craze has left a few horticultural legacies, one of which is the popular house plant the Boston fern, *Nephrolepis exaltata* 'Bostoniensis'. In 1894 the Philadelphia firm Robert Craig and Company shipped one hundred sword ferns (*N. exaltata*) to F. C. Becker of Boston, Massachusetts, who noticed that one of the batch was different, with broader feather-shaped fronds and a weeping rather than erect growth.[19] An alternative source attributes the discovery to the Soar Brothers, Miami nurserymen, who sent the sport up to Boston. The wild species is a Florida native and the plant needs frost-free growing conditions. Harry Ustler, a young clerk at the Springfield Floral Company, Illinois, realized that producing Boston ferns in Florida would earn better profits, and in 1911 he set up a nursery near Apopka, close to Orlando. The fern became an early mainstay of the Florida foliage and house-plant business. The business grew: by 1923 Apopkans were calling their town 'Fern City', and by 1927 Ustler was shipping out more than a million ferns a year. Business diversified into other foliage and pot-plant crops such that by the 1950s Fern City was the 'Indoor Foliage Capital of the World'. South Florida remains the world centre for the mass production of foliage plants.

While the Victorian craze for natural-history collecting kept going, there was a competing fascination for aquaria. Technology was established for the manufacture of Wardian Cases that could be adapted to aquaria, the relationship between aquatic plants and the oxygenation of water had been established, and the railways could carry the collecting hordes through the denuded fern sites to the new collecting grounds of the seaside. Robert Warrington was the publicist and initiator of this craze, and the early aquariums were called Warrington Cases, otherwise Aquatic Plant Cases or Parlour Aquaria. Soon cases were being designed to contain both aquatic habitats with fish and water plants and terrestrial areas with ferns and the occasional frog. This was soon followed by new public aquaria in places such as Sydenham (at the Crystal Palace), Brighton, Southport, Yarmouth, Manchester and Westminster. The Brighton Aquarium, still in existence, originally featured fern gardens and grottos in addition to tanks of fish or unfortunate porpoises, as did the lost aquaria.

The pioneering aquarist Takashi Amano (1954–2015) took the Victorian legacy of the enclosed world and, through an intimate understanding of aquatic plant physiology and a superb sense of design, created a series of aquatic landscapes that would have reduced Ward and Warrington to tears. Far beyond the grim algae-wrapped reality of most domestic aquaria, these are sophisticated distillations of underwater worlds where the plants are the dominant life form. Amano has had a huge influence on both amateur and commercial aquarists, and he established the International Aquatic Plants Layout Competition for Aquascaping.

The Wardian Case as a metaphor for the world has been explored by both artists and scientists. In the early 1970s the Argentinian artist Luis Fernando Benedit created the Phitotron, in effect a small green-house, or large Wardian Case, that explored ecological cycles. In 1972, at the Museum of Modern Art in New York, he installed a hydroponic greenhouse environment containing 70 tomato plants and 56 lettuce plants artificially supplied with light and a liquid fertilizer.[20] A closed world and a metaphor for environmental change.

You could argue that the atmospherically sealed Biosphere 2 experiment in the Arizona Desert is the largest Wardian Case to be constructed. However, in a reversal of the Victorian tabletop installation, eight people lived inside this case and looked out on to the wider world. Constructed between 1987 and 1991, it was built as a closed ecological system to allow the study of ecological and atmospheric processes. The structure covers 1.27 hectares (just over 3 ac), and its interior includes living quarters, an agricultural zone, dry and arid habitats, a 2,000-square-metre (21,500 sq. ft) rainforest and an 850-square-metre (9,100 sq. ft) ocean. Into this series of habitats a variety of plants and animals were introduced, including the eight people (the 'biospherians') mentioned above. While the two early experiments in total isolation and atmospheric self-sufficiency were controversial, Biosphere 2 continues under the management

overleaf: Biosphere 2, possibly the largest Wardian Case in the world, Tucson, Arizona.

of the University of Arizona in order to undertake research into ecological processes.[21]

The varied works of the artist and museologist Mark Dion often carry an echo of the Wardian Case and the associated Victorian traditions of specimen collection and display.[22] One such echo is Dion's Neukom Vivarium in Seattle, a crystal box to house a decaying trunk of a giant hemlock (*Tsuga*) tree. The Vivarium re-creates the original forest habitat through light and humidity, enabling the tree's original companions of ferns, mosses, fungi and soil to thrive and carry on the processes of decay. Shirley Hibberd would have approved.

New landscapes for indoor spaces are being built in the twenty-first century. Artists such as Giuseppe Licari are creating 'green rooms', areas of rolling turf and flowering trees within the sterile confines of the classic white space of a contemporary art gallery. This work is being taken to the next level by Baracco + Wright Architects, who installed a re-creation of the highly threatened Australian grasslands at the 2018 Venice Biennale. They used more than 10,000 plants comprising 65 different grassland species to represent the 1 per cent of the State of Victoria's original grasslands that have survived to the present day.[23]

We can see the early evolution of the vertical wall with the work of the Brazilian horticulturist, landscape architect and artist Roberto Burle Marx. Burle Marx had the Brazilian flora as his resource and the opportunity to work as an artist when Brazil was rediscovering its tropical identity. In an early garden, Guarita Park (1973–8) in Torres, southern Brazil, he used native bromeliads to create a vertical garden on a natural cliff. Working with architects such as Le Corbusier and Oscar Niemeyer, he developed designs for hanging gardens that used tropical plants. One of these was the Banco Safra project in São Paulo with Niemeyer, with characteristic plant columns and internal plant panels.[24]

The interior plantings of Patrick Blanc have released the fantasy world inside a Wardian case into our living spaces: Sky Team Lounge, Heathrow Airport, London (*above*), and Sofitel Palm Jumeirah, Dubai.

A modern take on the Wardian Case: Jamie North, *Inflection*, 2019, blast furnace slag, concrete and blown glass.

Marx adhered to no single professional channel, rather blending his knowledge of plants and ecology with his design instinct. In a similar polymathic vein, the tropical botanist Patrick Blanc has spearheaded a movement to bring large-scale tropical plantings into homes and indoor spaces such as offices, showrooms, museums and hotels.[25] Blanc has spent a lifetime studying tropical plants in the field. His inspirations range from the forests of Thailand to the flat-topped massifs of Venezuela, and through innovative horticulture he has created a series of extraordinary indoor installations. In a sense his work has grown from the aquarium and terrarium and expanded through and around buildings. He uses an exotic palette of plants, including tropical rheophytic (stream) species that are seldom seen in cultivation, in addition to familiar house-plant species such as *Caladium* and *Scindapsus*. Blanc has placed his visitors directly inside a romantic and diverse Wardian Case, where extraordinary plants cascade down walls, often animated by rivulets and wrapped in mist.

For decades the term 'house plant' has meant a plant in a pot, case or jar. Blanc and other innovators have developed the tools and the plant palette so that house plants can be grown at scale both in a house and, importantly, around a house.

five

The House of Plants

✿

The co-evolution between us and house plants is accelerating, a reflection of the complex changes in our society. The next stage of this co-evolution, while exhilarating, may not be without controversy. This chapter starts in a railway-station waiting room in Vienna, where monstera-inspired motifs have created an extraordinary temple to the exotic; meanders via Californian architecture of the 1950s where potted plants were key to the design of the new buildings; and ends as we look at innovative projects that are beginning to place our future within buildings moulded and wrapped by plants, algae and fungi – quite literally houses of plants. Linked to this new hybridity of construction is the blending of robotic functions with living plants and the application of new molecular tools in creating new plants. Strangely, as buildings become more organic, perhaps more natural, so our house plants may become less natural: genetic engineering and robotics will inevitably mould the plants we may want to bring home.

Once a set piece – a pot on a windowsill, a motif on the wall – the house plant is expanding in function to encompass living systems. The steady accumulation of horticultural expertise and the pressures of a changing world are pushing innovators and designers to look at how plants can improve our living spaces beyond the provision of ornament. While we know that the house plant provides potent support to our health and well-being, the next stage of our co-evolution will be dramatically different in both scale and scope. Environmental challenges are prompting questions about the future role of plants in crafting a sustainable, and above all liveable, urban biome. Just as the Wardian Case was a product of Victorian intellectual and commercial vigour, so today our relationship with the plants that enhance our living spaces is driven by twenty-first-century technology and creativity.

The waiting room in a Viennese railway station is a good place to start, and perhaps exemplifies that early relationship, plants as two-dimensional decorative and exotic motifs. The Austrian architect Otto Wagner loved the exotic and wrapped plant elements into both the exteriors and interiors of his buildings, notably the spirit-lifting exuberance of the Majolica House in Vienna. However, his most exotic creation was the interior of the royal waiting room for the Imperial Railway Station near Schönbrunn, the spiritual heart of aroid culture in Europe in 1905.[1] The waiting room's carpets and walls share a leitmotif of monstera leaves and aerial roots. What could be more appropriate for a royal court whose botanic gardens contributed so much to the science and horticulture of the tropical aroids? Wagner's design illustrates a long love affair with tropical plants that continues to grow in complexity and relevance from ornament and symbol to a mesh that supports life.

This love of the exotic decorative motif is remarkably persistent, and the big-leafed tropical plants function as shorthand for the tropical and exotic. In 1941 the textile company CW Stockwell asked the illustrator Albert Stockdale to crystallize the tropical ambience of the Bahamas into wallpaper.[2] The resulting Martinique range with a banana-leaf motif was launched the following year, and became famous through its exuberant adoption by the Beverly Hills Hotel. The range is still commercially available.

Big-leafed house plants, particularly the aroids such as *Philodendron* and monstera have become part of urban design and the public imagination. The monstera fascinated the artist Henri Matisse, who encountered it in the colonial gardens of subtropical North Africa, and it provided a suitably exotic backdrop for James Tissot's painting

Otto Wagner's Viennese railway station, decorated with monstera motifs.

In the Greenhouse (1869). They are part of the scenery of many tropical-themed movies, be it a frame from Alfred E. Green's *Copacabana* (1947), starring Carmen Miranda, or the backdrop for the profoundly dubious Tarzan films of the Johnny Weissmuller era.

The philodendron became the plant of choice for a generation of American and European modernist architects and designers. Many a beautiful open-plan house with wide plate-glass windows had

CW Stockwell Martinique® wallpaper, designed by Albert Stockdale.

House plants are a core part of contemporary interior design.

a potted philodendron, sometimes a monstera, as a focal point or masking the junction of windows. American architects such as Richard Neutra, A. Quincy Jones and Rodney Walker used them as counter-point to the stark lines of their buildings.[3] In Europe this approach was championed by the Danish architect Arne Jacobsen. These tropical plants, derived from the rainforest and favoured by the Victorians and Art Nouveau designers, had been adopted by a different generation for whom the bold leaves of the potted specimen contrast with the sleek design of the architecture. This approach was adopted by the Festival of Britain (1951), an event that pioneered the use of large containerized specimens in public spaces in the United Kingdom.

The traditional, largely ornamental role of house plants is broadening into the design and metabolism of buildings. We know that if plants are to improve the atmosphere of our homes, we must maintain large volumes of actively growing plants with associated architectural and engineering support. However, kick-started by Patrick Blanc's vision of green buildings and urban canyons, combined

Roy De Maistre, *Interior with Lamp*, 1953, oil on board.

with the recognition that we need to improve living conditions in the urban biome, we are seeing plants being used to bind rooms, wrap buildings and coil into the landscape, a truly three-dimensional relationship. Designers and architects are merging interior and exterior landscaping, blending engineering and horticulture, art and science. This approach is bold in its ambition, using large-scale planting to create new urban habitats that are stunning pieces of design that function as part of the urban environmental metabolism. Novel approaches are extending the role of house plants to encompass moss, fungi and algae as growing components of the house, and at the extreme pioneering edge of this hybrid zone is the development of the plant biohybrid, a merging of the living plant and micro-engineering.

Mervyn Peake's *Gormenghast* trilogy (1950–59) describes a great rambling edifice of a kingdom, wrapped in ancient trees and ivy, as much dank leaf mould as mortar and masonry. Today in Singapore some extraordinary buildings are demonstrating a regenerative version of Peake's plant-wrapped city, one that is altogether more sunny and exuberant than gloomy Gormenghast. This vision is developing

Barriers between room, structure and garden are broken down: Gohar Dashti, *Untitled*, from the series *Home*, 2017, installation photograph.

Tower 25, Nicosia, Cyprus, designed by Jean Nouvel, illustrates the blending of internal and external planting.

from the urgent needs of tropical Asia's megacities to support a densely packed population, to build skyscrapers that are a joy to live in and are able to ameliorate urban heat islands, slow the flow of storm water and reverse the loss of biodiversity. The Oasia Hotel Downtown designed by WOHA captures this philosophy: a thirty-storey hotel wrapped in a vast trellis that supports a range of tropical climbers, with trees growing from terraces and internal spaces shaded and cooled by the vegetation. A huge vertical habitat has been created, and in the plant mesh are many species we recognize as house plants. Another example is the Khoo Teck Puat Hospital in Singapore, which is wrapped by terraces of lush tropical vegetation.[4] We get a hint of this ongoing transformation in cities such as Miami or Hong Kong, where ornamental figs, brought in as decorative plants, germinate in the cracks in buildings and road bridges, suggesting a future Angkor Wat-style metropolitan labyrinth.

Such innovations are not restricted to the tropics, as we see from the spectacular Bosco Verticale in Milan, designed by Stefano Boeri. A new generation of skyscrapers is merging the idea of farm and apartment. The Austrian architecture studio Precht is developing a concept, the Farmhouse, for modular housing where residents produce their own food in vertical farms with both indoor and outdoor growing spaces.

Such projects are building a new role for plants as an integral component of the matrix and identity of the building. It will be fascinating to watch the ecology of these vertical ecosystems mature as species colonize the plantings, fungal associations diversify and the different species respond to sun, heat, water and nutrients over future decades. These will – if they are allowed to mature – become green wild-urban canyons, with an ecological trajectory we can only guess at.[5] Ironically, these new green canyons designed to support a sustainable future echo the street canyons of war-torn central Beirut, where among the piles of rubble and the growing thickets of ailanthus and

overleaf: The Park Royal, Singapore, designed by the WOHA practice, creates a modern city landscape where external plantings merge with internal ones.

fig trees you could spot house plants emerging from blasted balconies: a once potted *Phoenix roebelenii* arching out from a glassless window, a cluster of aloes and a spot of red pelargonium.

This biophilic approach is taken a stage further if the plantings in and on the building are explicitly designed as a habitat and conservation resource. In 2020 the sustainable design group Terreform ONE developed the concept for a building that would support the breeding of the monarch butterfly in New York City.[6] The Monarch Sanctuary comprises eight storeys of retail and office space. However, the purpose of the building, and its identity, is to serve as a breeding ground and sanctuary for the monarch butterfly (*Danaus plexippus*), and it is described as a 'new biome of coexistence for people, plants, and butterflies'. The urban populations of the monarch butterfly will be supported by open-air plantings of milkweed and nectar flowers on the roof, rear facade and terrace, and by partially enclosed breeding colonies in the atrium and street-side double-skin facade. Giant LED screens will provide views of the insects to the passing public. Interior structures will be grown from fungal mycelia, and sacs of photosynthesizing algae will help to purify the air and the building's waste water. Solar panels on the roof provide renewable energy to assist in powering the facilities.

The Monarch Sanctuary is exploiting different kingdoms of life, with its fungal structures and photosynthesizing algae. Our definition of the house plant, while decidedly loose already, will increasingly come to include other groups of life, such as fungi, mosses and algae. These are groups that in the next few years will cross into the home's internal living spaces to provide both decoration and life support.

The Bio-Intelligent Quotient in Hamburg is the first algae-powered building in the world.[7] The facade of this five-storey passive house contains tanks of aquatic microalgae, which as it grows is harvested and used as biofuel; in addition, the water tanks insulate the building and collect solar energy. The heated water is used directly

A house of plants: Bosco Verticale, Milan, designed by Boeri Studio.

for hot water and heating or stored underground using borehole heat exchangers. Future applications include cultivating the algae for food or fertilizer (ideally used for the landscaping inside and on the building), or filtering polluted air. After the algae in each tank reaches a certain stage of maturity, it is harvested and converted to biogas, which, if burned, can provide heat in the winter months.

A new generation of house plants? Algae being incorporated into domestic spaces as decoration, food source and fuel, installation by ecoLogic Studio at the 2021 Venice Architecture Biennale.

The next house-plant pioneer zone: the colonization of outside walls with plants, moss and algae.

The original algae for this project was collected from the nearby River Elbe, a 'wild' collection. Following the pattern of domestication demonstrated with house plants, we can expect a rapid domestication process with selections made for physiological efficiency and pigmentation for aesthetic impact. The interior designer Hyunseok An anticipated this and in 2019 designed a series of indoor algae farms to decorate a flat or home.[8] The wall-mounted geometric aquaria contain living colonies of algae that change colour as each colony matures. The French biochemist Pierre Calleja has created a prototype for a 'smog-eating' street lamp that uses bioluminescent algae to light streets while absorbing carbon dioxide and producing oxygen.[9]

Another photosynthetic life form, moss, is being used to clean the air of our living spaces. This ancient life form has the ability both to decorate our rooms and to clean the air (although many of the green moss walls being promoted for interior decoration are comprised of dead reindeer moss, a lichen, that has been stained various shades of green).[10] The moss absorbs large amounts of pollutants, particularly black carbon, which is both a potent greenhouse gas and a threat to human health.[11]

The use of a biological concrete mix has the potential to turn sterile, lifeless walls into living habitats through encouraging colonization by moss, lichen and ferns. Innovative work by the Universitat Politècnica de Catalunya (BarcelonaTech) in Spain has developed a concrete with a biological layer that collects and stores rainwater, providing a moist growing environment where microalgae, fungi, lichens and mosses can thrive.[12] Similarly, researchers at the Indian School of Design and Innovation in Mumbai have developed bricks made from a mix of soil, cement, charcoal and loofah fibres. The aim is to create a building that invites colonization by plants and insects on the outer surfaces.[13]

This is the start of a brand-new horticultural palette, the greening of the home walls to create new habitats through natural colonization, with the ecology of each wall varying according to aspect, prevailing humidity and air pollution. The next level of sophistication will involve selecting the blends of moss, algae and lichen to decorate a house wall; a whole new world of horticultural trendsetting and elitism awaits.

A glimpse of the future is provided by the New York design studio The Living, which created Hy-Fi in Queens in 2014, a set of three 13-metre-tall (43 ft) towers constructed from 10,000 bricks made of plant waste and fungal mycelia, a living and ultimately recyclable building.[14] After three months the towers were dismantled, the bricks composted and the resulting soil used in local community gardens.

Molecular science has become an influential and expansive field, and the next generation of house plants is likely to be developed by artists, engineers and horticulturists, using the extraordinarily potent tools of molecular genetics. The Brazilian American artist Eduardo Kac has stretched the definition of art and created new biological entities, transgenic art.[15] For his project *Natural History of the Enigma* (2009), he isolated a gene from his own genome, inserted it into the growing cells of a petunia plant, then had the plants cultured

Terence Conran interior design exhibition at Simpson's of Piccadilly, London, *c.* 1949–56, photograph by Nigel Henderson.

and grown into fully functioning specimens for exhibit. This everyday, fundamentally 'cheap and cheerful' bedding plant was now manufacturing a human protein. However, to the horror of some, Kac's hybrid has been created by an artist, not a scientist or horticulturist, and this new crafting will both disrupt and accelerate the older process of hybridization. In many respects Kac has repeated the eighteenth-century breeding experiments of Thomas Fairchild by transgressing 'sacred' boundaries, although while Fairchild may have feared for his soul, Kac appears to revel in the dilemma.

Kac's choice of plant, while influenced by the fact that petunia is an established 'lab rat' for research, also recognizes that this plant is a commonly traded commodity, planted by the million every summer, simultaneously banal and potentially beyond control. In 2015 a Finnish plant biologist, Teemu Teeri, spotted a summer planting of petunia containing bright orange flowers, and was reminded of some genetically engineered (transgenic) cultivars created thirty years before. His keen eye had indeed spotted a series of transgenic plants, and such plants are illegal in Europe. They had been inadvertently incorporated into breeding programmes and subsequently released into the horticultural trade. As a result, probably hundreds of thousands of illegal petunia plants were destroyed by 2017. The petunia is a perfect metaphor for the challenges facing us, with a long history in cultivation dating back to the 1830s, yet illustrating the risk and potential of the molecular age of plant breeding. While those transgenic petunias pose little or no known danger to the ecosystem or human health, the great 'transgenic petunia carnage of 2017' shows that plant-breeding programmes are profoundly leaky and can generate controversy.[16]

The other barrier that is being challenged is the merging of living and automated systems. Using a terrarium (the contemporary Wardian Case), the artist Saša Spačal has explored the potential for cross-kingdom communication – in the case of *7K: new life form* (2010), the communication between human, plant and fungal physiological processes.[17] The 7K, or Seventh Kingdom, to which Spačal refers is a new evolutionary kingdom, a kingdom of living matter expressing the ideas, tools and needs of humans. The Seventh Kingdom is already here with biohybrid robotics. The roots, stems, leaves and vascular circuitry of plants distribute chemical signals that regulate growth and function. Researchers have looked at these systems and made the comparison with electric circuits. For instance, a group of Swedish scientists have established functioning electrical circuits from living plant tissues.[18] Scientists and engineers are already exploring how to integrate plants and fungi into automated construction tasks and the physical frameworks of buildings.[19]

A team at the Massachusetts Institute of Technology (MIT), working under the research flag of plant nanobionics, is researching how to incorporate nanomaterials into growing plants to improve the efficiency of photosynthesis and provide new functions, such as monitoring pollutants.[20] One project suggests a future collaboration between the house-plant industry and MIT's plant nanobionics team: glow-in-the-dark plants, a superficially whimsical project that has profound implications. Using the naturally occurring enzyme system that gives glow-worms their glow (the luciferase system), the MIT team packaged the enzyme system into different nanoparticle carriers and inserted them into living plants. Future work may increase the amount of light and its duration, perhaps allowing one treatment to last the lifetime of the plant. There is no doubt what Harry Veitch would have done if given access to this technology – the Veitch catalogue would be boasting glowing bedside *Streptocarpus* plants and avenues of glow-in-the-dark monkey-puzzle trees.

MIT engineers are developing ways to track a plant's response to stress such as injury, infection and light damage, using sensors made of carbon nanotubes that are embedded in the plant tissue. These sensors report on levels of hydrogen peroxide, a signal for physiological stress within the leaves that promotes the repair of tissue and activates defence mechanisms against insects or fungi. Levels of hydrogen peroxide can be monitored by a camera attached to a tiny Raspberry Pi computer that can then send a message to a smartphone.[21] Now another stress-inducing app can be added to your phone – the house plants will be calling you.

In a similar vein to the MIT work, Rutgers University in New Jersey has developed another solution to dried-out and crispy house plants. Groups of FloraBorgs – effectively plant pots on wheels – roam the university corridors.[22] Each one is equipped to satisfy the environmental needs of the plant it contains, summarized as light and water. Solar panels on the side of the pot sense light, charge the battery and communicate with a small built-in computer that directs the plants towards the brightest source of light. A moisture detector in the pot alerts the computer and guides it to the nearest water fountain, where it signals the need to be watered to passing students. The Chinese company Vincross has created a six-legged crab-like robot to allow potted plants to follow the sun. Another variant, Elowan, created by the MIT Media Lab, taps directly into the plant's physiology with electrodes that respond to calcium gradients, and generates the electricity that communicates with the robotic mobile base.

In their collaborative project *Resurrecting the Sublime* (2019), the artist Dr Alexandra Daisy Ginsberg, the smell researcher Sissel Tolaas and a team from Ginkgo Bioworks have created a time machine.[23] The smell of an extinct relative of the hibiscus, *Hibiscadelphus*, last enjoyed on the slope of a Hawaiian mountain in the early twentieth century, has been resurrected. Taking DNA from herbarium samples of extinct plant species, the team used synthetic biology to resynthesize the gene sequences for scent. These sequences then guided the reconstruction of the fragrance. Is the next step to insert those gene sequences into a living house plant?

Our relationship with house plants represents in microcosm our relationship with the environment and wild biodiversity. We are fast developing tools that can create the sublime and the marvellous. We

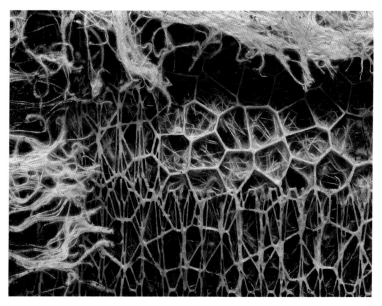

Grown textile roots system: Diana Scherer, *Interwoven* (*Exercises in Root System Domestication*), 2018, roots and soil.

Diana Scherer. *Nurture Studies*, 2012, soil, seed, photography.

can recreate the scent of a long-extinct plant, we can forge plants that glow in the dark, and we can plan for future buildings that are a matrix of living plant tissue and cement. We could be approaching a time of wonder and controversy.

Each house plant growing on a windowsill is a link to a bigger world. They embody the history of botanical exploration and the growth of plant breeding and genetics. Over the last four hundred years we have forged what may be the world's most diverse experiment in plant domestication, transforming the anatomy and physiology of wild plants through art and technology. Novelty still rules the house-plant business, and novelty encourages the crossing of boundaries, whether it be making a simple cross between two flowers, using radiation to generate a new African violet or inserting a human gene into a petunia. This co-evolution with the plants that we invite into our homes will keep shifting as ecology, ethics and the economy change.

six

Wild and Endangered Relatives

❦

I t is difficult to reconcile a shop-bought and cellophane-wrapped *Echeveria* or African violet with their wild ancestors. Domesticated house plants can be as different from their antecedents as a supermarket chicken is from a wild jungle fowl. House plants that are propagated and sold by the hundreds of thousands, even millions, may be derived from wild populations that are extinct or facing extinction. Such house plants carry echoes of lost landscapes now erased by goat, fire or plough.

History tells us that as new plant groups are discovered, so some horticulturists will demand wild specimens and render those wild populations vulnerable to predatory commercial collectors. The nineteenth- and twentieth-century commercial collectors of orchids and succulents exemplified this buccaneering approach, protecting their livelihoods through death threats, thuggery, deception and lies, the sabotage of consignments and even the destruction of wild stock to ensure high prices. Collections by the tonne were shipped back to European and American markets, often with very high levels of mortality. Sadly, some individuals will still pay a premium for wild and illegal plants.[1] A dangerous allure attaches to a scratched and scarred wild plant, even though commercially propagated plants are easily available and will perform better in cultivation.[2]

Until relatively recently, large wild harvests of bulbs, cacti and succulents, carnivorous plants and bromeliads were sold in the United States and Europe for the amateur market. The introduction of CITES

The Christmas cactus, *Schlumbergera*, an epiphytic cactus originating from the threatened montane forests near Rio de Janeiro, Brazil.

(the Convention on the International Trade in Endangered Species) as international law has had a hugely beneficial impact on regulating this damaging trade and encouraging the horticultural production of desirable plants.[3] Today the trade in garden favourites such as snowdrops (*Galanthus* spp.) is almost entirely based on cultivated plants. The vast majority of house plants sold in shops are commercially grown and no longer rely on the pillaging of wild stocks, however it is always worth checking whether a specimen is wild collected, especially with high-value plants such as succulents.

Many of our popular house plants come from areas identified as 'biodiversity hotspots', tiny, magical patches of the Earth's surface characterized by both high levels of uniqueness (endemism) and high levels of habitat destruction.[4] For instance, the Atlantic Forest Biodiversity Hotspot of southern Brazil is the source of the domestic gloxinia (*Sinningia speciosa*) and the Christmas cactus (*Schlumbergera* spp.). African violets are a case in point; they are endemic to a series of forested mountains that stretch between Tanzania and Kenya, the Eastern Arc Mountains, a biodiversity hotspot that sits above the dry lowlands and provides refuge for extraordinary numbers of endemic species of bird, reptile, mammal and plant.[5] The first taxonomic botanist to study the wild *Saintpaulia* species was Bill Burtt of the Royal Botanic Garden, Edinburgh; working with a relatively small number of herbarium specimens, he recognized a total of twenty species.[6] Since his initial work, the number of field collections has increased, providing much-improved survey data and a better understanding of the plant's natural variation and ecology. This has been combined with molecular phylogenetics, using genetics to map the evolution of a plant group, to provide a more detailed understanding of the biology and conservation of African violets. The latest study has recognized six wild species, and a further two were recently identified in the Uluguru Mountains of Tanzania.[7]

The wild ancestor of the cultivated African violet, *Saintpaulia ionantha*, with the widest geographical and altitudinal range, is not immediately threatened with extinction but has suffered significant habitat loss since the early twentieth century. Related species are much closer to extinction. For instance, *S. teitensis* is restricted to a tiny patch of montane forest in the Taita Hills, southern Kenya, where the total wild population is concentrated within less than 1 square kilometre

One of the most threatened of the wild African violets, *Saintpaulia teitensis*, survives in a tiny patch of forest in the Taita Hills, Kenya.

(just over ¼ sq. mi.).[8] Another species, *S. ulugurensis* from the Uluguru Mountains, Tanzania, is thought to have a wild population of fewer than thirty and to survive on a single site covering just 5 square metres (54 sq. ft).[9]

All the wild *Saintpaulia* species are at risk of extinction from habitat loss that is driven by the rapidly increasing populations of Kenya and Tanzania, and the desire for new agricultural land, wood, charcoal and limestone for cement factories. The trajectories to extinction are likely to accelerate further as a result of the impact of climate change on the region's temperatures and rainfall.[10] This begs the question of what the relationship is between commercial trade in a plant and the conservation of the ancestral and related wild diversity. Unlike other traded ornamental plants, the African violets have not been subject to massive and destructive wild collecting (as happened with air plants, *Tillandsia*, in Central America and succulents in Madagascar, for example), but wild genes have been used for commercial breeding. For instance, *S. grotei* (now recognized as a subspecies of *S. ionantha*) was crossed with *S. ionantha* stock in the 1950s to breed trailing cultivars.

Following recent phylogenetic research, the genus *Saintpaulia* has been found to be part of *Streptocarpus*, so it has, in botanical parlance, been 'sunk' into *Streptocarpus*.[11] This means that the name 'saintpaulia' will probably be used as a vernacular or horticultural name, along with 'African violet', but in scientific terms *Saintpaulia ionantha* is now *Streptocarpus ionanthus*. It is sad, in a way, that history has been traded for botanical accuracy. Perhaps more seriously, conservationists use endemic genera as part of a conservation argument that the combination of genetic uniqueness and restricted geographical range gives added conservation priority to a plant and its habitat. *Saintpaulia* is a lovely and potent flagship for the highly threatened and magical forests of the Usambara; now, as *Streptocarpus*, that role has perhaps been diminished.

Two charming house plants are derived from the mysterious island of Socotra, off the coast of Yemen, and part of the Horn of Africa Biodiversity Hotspot. These plants, collected from an almost

Exacum affine, the Persian violet, collected from the arid island of Socotra.

mythical island more than 140 years ago, have contributed genes to some of the most popular decorative plants sold in the garden centres of Europe and North America. The first, the pretty Persian violet, *Exacum affine*, was discovered on the island in 1880 by Isaac Bayley Balfour, then Regius Professor of Botany at Glasgow University and later Keeper of the Royal Botanic Garden, Edinburgh, and has become a popular house plant.

The second plant found on the same expedition, *Begonia socotrana*, was subsequently used to create the exuberant winter-flowering begonias. It was originally thought to be highly endangered, but recent fieldwork has shown this species to be relatively abundant on Socotra, growing in sheltered north-facing rock crevices.[12] Plants of *B. socotrana* were sent to the Royal Botanic Gardens, Kew, by Balfour and flowered in the winter of 1880; from there plants were sent to the Veitch Nurseries, which started distributing plants in 1882. John Heal of the Veitch nursery saw the potential of this pink-flowered species and made the first cross with a Mexican species, *B. insignis*; the resulting cultivar, 'Autumn Rose', flowered in 1882. The next – appropriately enough, called 'John Heal' – was a cross with *B.* 'Viscountess Doneraile', a hybrid made by another Veitch breeder, the remarkable John Seden.[13]

These initial crosses paved the way for an ongoing series of new cultivars that are popular house plants, such as the Rieger hybrids.

On the other side of the world another tropical archipelago, the Hawaiian islands, is the home of a relatively new addition to the house-plant pantheon, *Brighamia insignis*, known by the beautiful Hawaiian names *ālula*, *'ōlulu* and *pua ala*, in contrast to the somewhat

One of the last known wild plants of *Brighamia insignis* on the north shore of Kaua'i, Hawaii.

more prosaic Western 'volcano palm' or 'Vulcan palm', or the frankly demeaning 'cabbage on a stick'.[14] The *ālula* is a spectacular, some would say bizarre, member of the campanula family that grows into a tall-trunked plant with an admittedly cabbage-like crown of leaves from which beautiful, highly scented yellow-white flowers emerge. It was once abundant on the cliffs and escarpments of Kaua'i and Ni'ihau, but has not been seen in the wild since 2012. Fortunately, plants and seed were collected by field conservationists from the National Tropical Botanical Garden (NTBG) on Kaua'i, and it has been successfully established in cultivation. In the 1980s seed was sent from NTBG to botanic gardens around the world, and plants eventually entered the house-plant trade.[15] A selected cultivar, 'Kirsten', is available in the European market; it can grow into a spectacular swollen-trunked pachycaul, but it does function as a potent magnet for red spider mites.[16]

While the endemic plants of Hawaii were becoming increasingly restricted to mountain refuges above the expanding sugar-cane fields, a group of tropical horticulturists were hybridizing hibiscus, a plant that is powerfully symbolic of the tropical sugar islands. The tropical hibiscus, *Hibiscus rosa-sinensis*, while primarily a shrub of tropical gardens, is a popular house plant and has an extraordinary and obscure history in cultivation. The original species was widely human-dispersed through Asia and Polynesia long before European contact, and probably represents an ancient domesticate with a long history of hybridization whose origins have yet to be confirmed.[17]

The early Hawaiian breeders, along with colleagues on other sugar islands such as Sri Lanka, Fiji and Mauritius, crossed some of the most endangered and mysterious hibiscus species to create ornamental hybrids that still exist today. The combination of their gardening interest and a network of botanic gardens and agricultural stations created a unique opportunity for breeding the tropical hibiscus. As early as the 1820s Charles Telfair, an Irish botanist then resident in Mauritius, was crossing *H. rosa-sinensis* with endemic hibiscus from the Mascarene Islands, and sending the resulting hybrids

Part of the breeding 'herd' of *Brighamia insignis* managed by the National Tropical Botanical Garden, Hawaii.

back to the nursery of Robert Barclay in England. In Hawaii the hybridizing of *Hibiscus* can be traced back to the 1870s, when the Hon. Archibald Scott Cleghorn, Governor of Hawaii, developed twelve new cultivars; another breeder was John Adams Kuakini Cummins, Hawaiian Minister of Foreign Affairs (1890–91), who crossed the Pacific *H. cooperi* with the African *H. schizopetalus*. However, breeding in Hawaii really took off with the emergence of the plantation classes,

A.B.del.J.N.Fitch.Lith.

Vincent Brooks Day & Son Imp.

Hibiscus schizopetalus, native to a few sites on the east coast of Africa, is one of the wild species that has contributed to the garden hibiscus. Illustration from *Curtis's Botanical Magazine*, vol. CVI (1880).

The tropical garden hibiscus, a complex artificial hybrid containing the genes of some of the world's rarest plants.

a network of politically well-connected and affluent families with a love of horticulture and natural history. From the early twentieth century they crossed *H. rosa-sinensis* with the endemic, fragrant Hawaiian hibiscus species (including the beautiful *H. waimeae*) and related species from East Africa (*H. schizopetalus*), the Mascarenes (probably *H. boryanus*, *H. liliiflorus* and others) and the South Pacific (including the newly discovered *H. macverryi*, then known as 'Mrs Hassinger', and the mysterious *H. cooperi*).[18] One of the species listed in those early breeding programmes, *H. liliiflorus*, is endemic to the tiny Mascarene island of Rodrigues, where the wild population was at one point down to three known individuals; conservation work by the Mauritian Wildlife Foundation has secured its survival. This beautiful red-flowered plant is the only hibiscus known to be pollinated by fruit bats.

Recent fieldwork in Fiji has helped to resolve the question of the identity of some of the wild species used in breeding with *H. rosa-sinensis*.[19] In 1860 a new species of *Hibiscus* was discovered in Fiji, *H. storckii*, but it was not seen again until 2016. During fieldwork by Lex Thomson, undertaken to rediscover this species, a further three new endemic species were found, *H. bennettii*, *H. bragliae* and *H. macverryi*, all highly endangered. While no evidence was found that *H. storckii* has contributed to hibiscus breeding, it was discovered that material of all three 'new' species had been distributed from Fiji under various names and used for breeding in Hawaii and Florida. Accordingly,

The highly endangered *Hibiscus liliiflorus*, endemic to the Indian Ocean island of Rodrigues.

the genes from some of the world's most endangered plants have contributed to the tropical hibiscus of our conservatories and gardens.

The succulent plants have historically been hit hard by unsustainable collecting to supply the horticultural trade. Around 30 per cent of all cactus species are threatened with extinction, and around half of these are threatened by the horticultural trade and illegal collecting.[20] However, a recent analysis of cactus seed in the commercial trade found that only a tiny proportion could be attributed to wild origins.[21] The golden barrel cactus, *Echinocactus grusonii*, from Mexico is propagated in cultivation by the million, and is a common but probably short-lived house plant in northern climates, yet is one of the world's most threatened cacti. Only two wild populations are known. The first to be discovered was at Querétaro in Mexico and has been badly damaged by illegal collecting and loss of habitat caused by the construction of a dam; a second and larger wild population was more

Hibiscus macverryi from Fiji contributed to the breeding of the domestic hibiscus, although the species was only recently discovered in the wild.

The golden barrel cactus, highly endangered in the wild but widely grown as an ornamental.

Commercial propagation of the highly endangered golden barrel cactus.

recently discovered in Zacatecas.[22] A massive salvage operation was mounted by Mexican botanists to rescue plants of the golden barrel cactus and other succulents from the Querétaro site; a total of 48,000 succulent plants, including golden barrel cacti, were relocated to safe sites above the new water level or taken to Mexican botanic gardens to create cultivated breeding populations.[23]

Another popular succulent house plant originating in Mexico is the ponytail palm, *Beaucarnea recurvata*, which is restricted to the semi-arid Tehuacán Valley. There it makes a large multi-stemmed succulent that can reach 10 metres (33 ft) in height, whereas as a house plant it usually forms a gangly, often mealybug-spattered 1–2 metres (5–6½ ft). This species is propagated by the hundreds of thousands in nurseries in the Canary Islands, California and Thailand. In the wild in Mexico it is close to extinction, categorized as Critically Endangered by the International Union for Conservation of Nature Species Survival Commission.[1] The illegal collecting of plants, urban expansion and over-grazing of the plant's arid bushland habitat have reduced wild populations by more than 80 per cent. Wild plants have been collected and established in commercial nurseries for use in both Mexico and overseas. However, the vast scale and efficiency of horticultural production mean that specimens purchased as house plants outside Mexico will likely have been legally grown in nurseries.

The Cape Biodiversity Hotspot of southern Africa holds one of the planet's most spectacular botanical assemblages. Within the hotspot's 90,000 square kilometres (35,000 sq. mi.) are some 9,000 species, of which 70 per cent are endemic, constituting an extraordinary 25 per cent of all African plant species. Many of those species have become much-loved house plants, including pelargonium, freesia, streptocarpus, spider plants, clivia and the Cape heaths. After the great horticultural passion for Cape heaths faded in the early nineteenth century and new plant passions were chased, the collections held by botanic gardens, commercial nurseries and private individuals declined rapidly. A few species lingered and could still be found in the occasional UK nursery, for instance the winter-flowering *Erica gracilis*. One such survivor is the verticillate heath, *E. verticillata*. Once cultivated and available in nurseries in the United Kingdom, it probably became extinct in the wild in the mid-twentieth century. This plant has escaped 'living dead' status, whereby botanic gardens harbour the last individuals of species now extinct in the wild, often

A mature ponytail palm from Mexico, *Beaucarnea recurvata*, in Miami.

with little hope of returning to the wild. Examples of the 'living dead' are *Sophora toromiro* from Rapa Nui, *Kokia cookei* from Hawaii and the venerable cycad *Encephalartos woodii* from South Africa, which survives as a single male clone.[25]

The Cape heaths, *Erica*, are a spectacular component of the Cape flora, with 680 endemic species out of a global total of 840 *Erica* species. With their large flowers, spectacular colours and winter-flowering habit they became the focal point of one of those strange periods of short-lived and collective horticultural passion.[26] Between 1787 and 1795 the Kew collections received seed of 86 *Erica* species from their collector Francis Masson. This inventory grew with new introductions, until by the end of the eighteenth century around 150 species were in cultivation. Europe was firmly gripped by 'Cape Heath Fever'. In 1826 the Scottish plant collector and professor of Greek at the University of Edinburgh George Dunbar could boast 344 *Erica* species growing in his Heath House. The undoubted authority was William McNab of the Royal Botanic Garden, Edinburgh, author of *A Treatise on the Propagation, Cultivation and General Treatment of Cape Heaths* (1832).

As the new species entered cultivation, so horticulturists started breeding new hybrids. Europe's glasshouses and nurseries were now full of a spectacular and chaotic mix of new species and hybrids with undoubtedly a large proportion of wrongly labelled plants. As with other such crazes, the collapse was quick; a susceptibility to rotting in winter and the ascendancy of new crazes, including Cape bulbs, caused many of the *Erica* collections to be discarded. Joseph Hooker commented in 1874 that 'the best collections of the present are ghosts of the once glorious ericeta of Woburn, Edinburgh, Glasgow and Kew.'

However, some plants of *E. verticillata* survived in cultivation. In 1786 Emperor Joseph II of Austria sent two horticulturists, Franz Boos and Georg Scholl, to the tropics to collect plants for the royal collections. The initial consignment, brought back in 1787 by Boos, included ten chests of plants, two live zebras, eleven monkeys and 250 birds. Scholl stayed in South Africa for a further fourteen years and sent shipments of seed and plants back to Austria, including seeds of *E. verticillata*. The resulting plants survived the decline of the European *Erica* collections and, most importantly, the Second World War. The two world wars had a catastrophic impact on European plant collections: horticultural staff went away to fight and many never came home, shortages of coal meant that glasshouses were not heated, and gardens were turned over to food production. Some collections were bombed, burned, raided for edibles or looted. However, despite the extreme privations of war-torn Vienna, miraculously the verticillate heath survived.

In the 1980s the South African horticulturist Deon Kotze began the search for the species, a quest that would expand from the local fynbos habitat and gardens of South Africa to the nurseries and glasshouses of Europe and North America. First Kotze scoured South Africa, locating plants in cultivation at the Protea Park in Pretoria. Cuttings were collected from the last surviving individual from an original population of three. This clone was named 'African Phoenix'. A plant from the Royal Botanic Gardens at Kew was located and subsequently confirmed as a sterile hybrid. With the keen eye and instinct of an expert horticulturist, Adonis Adonis found a plant of the verticillate heath growing in a clearing on the edge of the Kirstenbosch Garden, a survivor from an old collection of *Erica* species, probably a seedling from a collection made some seventy years earlier. Two clones have been derived from this source, named 'Adonis' to honour the rediscoverer and 'Louisa Bolus' after the original collector in 1917.

During this search the *Erica* expert Dr Edward George Hudson Oliver remembered seeing a plant in the Schönbrunn Palace collection in Austria. Plants were subsequently repatriated to South Africa and a fourth clone added to the founder population, clone 'Belvedere'. These plants had grown in Vienna since the original Boos and Scholl collections and survived the catastrophic bombing and lack of fuel during the Second World War. Additional plants were found in the United States (Monrovia Nursery in California) and the United Kingdom (Tresco Abbey Gardens in the Scilly Isles). In total eight

HEACOCK'S KENTIAS

Joseph Heacock Company
WYNCOTE, PENNSYLVANIA

One of the Palm Houses at Wyncote—Hundreds of "Palms that Please"

Catalogue for Heacock's Kentia palms, grown in Wyncote, Pennsylvania, with the seed
likely imported from Lord Howe Island, Australia, 1912–13.

clones were verified as the founder stock for the recovery of the
species and cultivated at Kirstenbosch National Botanical Garden,
Cape Town.

The first trial reintroduction was attempted in 1994 at Rondevlei
Nature Reserve, an area of sand-plain fynbos within Table Mountain

Park. This helped to identify the best habitat for the species and confirmed the presence of wild pollinators. Subsequently the plants have produced seed in the wild and the resulting seedlings have been planted at the nearby Bottom Road Sanctuary. Specimens were also planted at Kenilworth Racecourse, Cape Town, in 2005, and seedlings subsequently appeared on the site. An additional planting has established a population at the Tokai sand-plain fynbos reserve.[27]

During Victorian times a few tough plant species were found to survive the toxic fug of a household, and the infamous aspidistra was one such survivor. The other was the Kentia palm, or parlour palm (*Howea forsteriana*), native to Lord Howe Island in the South Pacific. From the 1880s, after the collapse of the whaling industry, Kentia palm seeds became increasingly important as an export crop to supply northern nurseries and the house-plant trade. It was pre-adapted to the cool temperatures of the Victorian parlour or conservatory, could tolerate poor air quality and low light, and was available as seed in dependable numbers that allowed mass production by nurseries: 'No plants are more easily grown and none are more tenacious to life than the palm, enduring alike dust and the hard knocks that house plants are apt to receive, the cold from open windows and the unnatural heat from furnaces and from gas.'[28]

This palm graced the palm courts of innumerable hotels, resorts and luxury liners (including the *Titanic*), and it provided the garnish to thousands of stiff and unsmiling Victorian family portraits.[29] The harvest of seed has been managed commercially since 1906, when the Lord Howe Island Kentia Palm Nursery was established. Until the 1980s only seeds were exported, but since then seedlings have also been sold. Seed is gathered from both wild plants and cultivated seed orchards, and shipped as bare-rooted seedlings, with profits invested in island conservation. Lord Howe Island exports about 375,000 seedlings a year; by comparison, in 1928 some 13,666,500 seeds were harvested and exported to Europe and America. The island nursery now faces increasing competition from growers in South Africa, Sri Lanka and California.[30]

Traditionally, the house-plant trade has brought plants from the wild into cultivation. We are now seeing a reversal of that flow. Sometimes house plants escape the veranda or home and become established away from their natural origins, and some of these have the potential to cause serious ecological damage. In parts of South Africa and Hawaii *Kalanchoe* hybrids originating in Madagascar are causing problems; elsewhere various tropical tradescantias are colonizing new habitats. Perhaps the most spectacular case of house

The felt bush, *Kalanchoe beharensis*, cultivar 'Fang'; the wild species comes from the threatened dry bushlands of Madagascar.

plants gone wild is featured regularly in films, a 'jungly' backdrop to everything from *Raiders of the Lost Ark* to *Jurassic Park*: the weird synthetic lowland forests of Hawaii.

These are termed 'novel' forests, a strange blend of exotic introduced plants that are forging a new ecology after the almost complete loss of the original forests. The vegetation looks like jungle, a facsimile of our vision of a rainforest, but biologically it is virtually sterile – it lacks the intricate set of ecological processes that defined the old Hawaiian forests. No one is quite sure how these forests will change over time. The hope is that more native species will colonize, but what is certain is that the big-leafed climbing aroids that have symbolized the tropics and graced houses throughout the world are now permanent denizens of this area. Our dreams of the benign jungle have become real in Hawaii, where woodland and palm thicket are draped in vines of pothos, *Epipremnum pinnatum*, monstera and philodendron.

For the conservationists working on the front line, the resources made available for commercial house-plant production must seem both dizzying and somewhat cruel. They are in fact two very different worlds. The conservation front line is often characterized by teams working to save species against competing fundamentals such as farming and charcoal harvesting while working with obscenely small budgets; in contrast, the house-plant world is based on discretionary spending by affluent markets for exotic novelties. Yet it seems impossible to comprehend that in a few decades we could witness the loss of the wild African violets from their mountain refuges while the domesticated forms exist as elaborately bred commensals in northern houses, dependent upon us for their survival.

We cannot rebuild the lost populations of threatened plants from our windowsills and commercial nurseries. Most house plants are clonally or vegetatively propagated, so all plants of any particular species or cultivar are likely to be virtually genetically identical or have been subject to heavy selection, rendering them unsuited to a wild existence. However, there are indirect means that allow house plants to play a potentially powerful role in conservation. We love house plants and they are a direct link to the world of wild biodiversity – an African violet is an envoy, a flagship, for the forests of Tanzania and Kenya. Yet this emotional link has never been successfully exploited. Is it time for a conservation tariff to be added to the purchase price of African violets, to fund the field conservation of the species and their forests?

Conclusion
New Worlds

🌿

In the science-fiction film *Silent Running* (1972), a lone botanist, Freeman Lowell, nurtures the last – clearly doomed – samples of plants rescued from Earth. These fragments of a lost world are a mix of big-leafed tropical plants, including the ubiquitous monstera, growing in a spaceship resembling a Richard Buckminster Fuller dome. A shelf of house plants, often including those same species from *Silent Running*, also testifies to a lost world. The world where the African violets were first collected has gone forever. In the 1890s, when Baron Walter von Saint Paul-Illaire sent seeds from the Usambara Mountains of Tanzania to his father in Germany, the world population stood at 1.6 billion people; it is now closer to 8 billion.[1] Over the same period atmospheric carbon dioxide has risen from 294 parts per million to around 415 parts per million.[2] In the last 130 years the rich forests of the Tanzanian mountains have been shredded into scattered fragments and the extinction process continues to accelerate.

House plants have proved themselves as a permanent part of our domestic lives, part of our multi-species households, with each generation of breeders moulding the plants to suit the taste of an ever-shifting market. Perhaps more than ever they are part of the infrastructure of well-being: they are commensal organisms that enrich our lives, becoming ever more deeply intertwined with the culture and metabolism of urban life. They are also the symbols, the totems, that indicate how society responds to global challenges.

The Usambara Mountains of Tanzania, where the first African violets were collected from the wild in the early 1900s.

To date, food has provided the emotional and ethical testing ground for opinions on genetically modified organisms and synthetic biology as a whole.[3] A future test will be whether we welcome genetically modified house plants into our homes. House plants with modified genomes are being produced, but so far public opinion is against them. The tools for this technology are increasingly available and costs are shrinking, so the opportunities for plant breeders will become ever more tempting. This powerful technology, as with previous technological breakthroughs, will undoubtedly present risks, and drive the production of tacky plant novelties (glow-in-the-dark aquarium fish, GloFish, are already available in pet shops in the United States), but it will also open the door to utilitarian opportunity. For instance, genetic modification can be used to improve the ability of house plants to absorb toxic pollutants.[4]

The carbon costs of the house-plant trade are profoundly unsustainable. Current house-plant production depends on a fossil-fuel-based economy of cheap plastics, cheap labour and cheap transport

– a global network of temperature-controlled trucks, air freight and shipping containers that links growers with markets in Asia, Europe and North America. Plant production, once based on one location with a single nursery that would take a cutting or seedling to final market-ready plant, has been replaced by a global production system that tracks cheaper labour and warmer growing conditions.

The indignity of domestication: rubber plant on the move, Vienna, 1954, photograph by Franz Hubmann.

There is an urgent need to calibrate the economic opportunities of globalization against the sustainability, equity and particularly the carbon budget of production.

One important topic is the growing medium, specifically the dependence of commercial plant production on peat. Nearly 3 million

INDIA-RUBBER PLANT
(FICUS ELASTICA)

India rubber plant, a popular house plant and once a source of a wild collected rubber. Illustration from Edward Step, *Favourite Flowers of Garden and Greenhouse*, vol. III (1897).

The India rubber tree as it should be, a giant forest tree whose roots have been braided together to form a living bridge in Meghalaya, India.

cubic metres of peat are sold for horticultural use in the UK for containerized plant production, including house plants.[5] Peat is dug from peatlands that are invaluable natural assets: they support unique assemblages of species and act as huge water sponges that protect communities from flooding, and they hold huge volumes of carbon that have accumulated over centuries and are released when bogs are harvested, so accelerating the ongoing climate crisis. The race is on to develop alternatives, with research groups working on treated manure from farms, coir, bio-char and vermi-compost, among others. In the very near future, the horticultural use of peat will be seen as an obscene anachronism.

The phylogenetic definition of a house plant will shift. Within the photosynthetic realms algae and moss will become increasingly used, both as interior decorations and as living membranes on the exteriors of buildings. Living colonies of fungi will be sold as dynamic living displays and perhaps as decorative processors of household waste.

Perhaps most importantly, each house plant, whether glorious or etiolated, will be an envoy for a wild and wondrous world that is fast being lost. A world where the living aerial roots of *Ficus elastica* are braided into bridges that cross deep ravines in northeastern India, where black rhino search out the succulent leaves of *Sansevieria* in arid African bushlands, where tropical butterflies pollinate *Clivia* lilies, and where pots of *Dieffenbachia* protect Amazonian households from evil.[6]

Timeline

❧

1500 BCE	Pharaoh Hatshepsut sends a plant-collecting expedition from ancient Egypt to the Land of Punt (probably Somaliland)
1608 CE	Sir Hugh Platt publishes his gardening manual *Floraes Paradise*
1630s	John Tradescant the Elder grows *Pelargonium triste*, starting the British love affair with 'geraniums'
1646	Charles Plumier, botanist and early aficionado of the big-leafed aroids, is born
1720	Thomas Fairchild exhibits his hybrid 'mule' at the Royal Society, London
1722	Thomas Fairchild publishes *The City Gardener*
1755–1817	Nikolaus Joseph von Jacquin establishes the aroid legacy of Vienna
1767	Philibert Commerson collects the first *Caladium* in Brazil
1799	Arthur Johnson of Yorkshire makes the first *Hippeastrum* artificial hybrid
1815	The wild collection of gloxinia, *Sinningia speciosa*, in Brazil
1828	Joel Roberts Poinsett finds a poinsettia plant, *Euphorbia pulcherrima*, in Mexico

1829	Nathaniel Ward notices ferns germinating in a sealed glass container
1833	A trial consignment of plants is sent in Wardian Cases from the United Kingdom to Australia
1840s	Frederik Liebmann and Józef Warszewicz Ritter von Rawicz introduce *Monstera deliciosa* into cultivation
1853	The first artificial hybrid orchid, *Calanthe* × *dominyi*, is produced by the Veitch Nurseries in England; it flowers in 1856
1880	Isaac Bayley Balfour collects *Exacum affine* and *Begonia socotrana* from Socotra
1891	Baron Walter von Saint Paul-Illaire sends African violet seed from Tanzania to his father in Germany
1893	Dr Henry Nehrling buys a collection of Brazilian *Caladium* at the Chicago World's Fair, starting the United States' caladium industry (worth approximately $12 million today)
1893/4	George Stumpp of New York imports the first African violets into the United States
1894	The Boston fern, a sport of *Nephrolepis exaltata*, is spotted by F. C. Becker of Boston, Massachusetts
1905	Otto Wagner designs a monstera-themed waiting room for the Imperial Railway Station, Vienna
1906	The Lord Howe Island Kentia Palm Nursery is founded
1909	Noël Bernard and Hans Burgeff discover that orchid seeds germinate with a symbiotic fungi
1946	The first African Violet Show is held in Atlanta, Georgia
1951	The Festival of Britain opens, introducing the use of house plants as part of contemporary design in the United Kingdom

1951	The liquid house-plant fertilizer Baby Bio is launched in the United Kingdom
1952	The term 'house plant' is coined by Thomas Rochford
1961	The first edition of Dr D. G. Hessayon's *Be Your Own House Plant Expert* is published
1970	Longwood Gardens (Pennsylvania) and the U.S. Department of Agriculture mount an expedition to New Guinea to collect *Impatiens* (busy Lizzie) breeding material
1973–8	Roberto Burle Marx starts the tropical vertical gardening style in Guarita Park, Torres, Brazil
1983	The first transgenic plant, a tobacco, is bred by Monsanto
1989	A NASA study on indoor plants and air quality is published
1994	Reintroductions of *Erica verticillata* are started in South Africa after the rediscovery of plants in cultivation
1998	Pioneering tropical green wall installation is carried out by Patrick Blanc in Genoa
2009	A transgenic petunia (called Edunia) with a human gene is created by the artist Eduardo Kac
2012	*Brighamia insignis* is seen in the wild for the last time
2014	A MIT team is working on 'bionic plants', including glow-in-the-dark specimens
2014	Sales of foliage plants for indoor and patio use in the United States are worth $747 million
2016	*Hibiscus storckii* is rediscovered in Fiji 156 years after its discovery in 1860
2017	The great 'transgenic petunia carnage', when illegal transgenic petunia hybrids are destroyed after illegally entering European trade

2018	Baracco + Wright Architects install the indoor Australian grassland exhibit at the Venice Biennale
2019	FloraBorgs trundle through the corridors of Rutgers University, New Jersey
2019	The scent of the extinct *Hibiscadelphus* is recreated by Dr Alexandra Daisy Ginsberg
2019	The Royal Horticultural Society in the United Kingdom announces competitive classes for house plants at future Chelsea Flower Shows
2020	Terreform designs a New York office building as a refuge for the monarch butterfly
2021	EcoLogic Design Studio exhibit algae culture as part of urban living systems at the Venice Biennale
2050	Nearly 70 per cent of the world's population will live in cities

References

Introduction: Plants of the Indoor Biome

1 Hugh Findlay, *House Plants: Their Care and Culture* (New York, 1916), p. 1.
2 Edward O. Wilson, *The Diversity of Life* (Cambridge, MA, 1992), p. 350.
3 Mea Allan, *Tom's Weeds: The Story of the Rochfords and Their House Plants* (London, 1970), p. 125.
4 David Gerald Hessayon, *Be Your Own House Plant Expert* (London, 1961).
5 Xi Liu et al., 'Inside 50,000 Living Rooms: An Assessment of Global Residential Ornamentation Using Transfer Learning', *EPJ Data Science*, VIII/4 (2019).
6 Harold Koopowitz, *Clivias* (Seattle, WA, 2002), p. 174.
7 Lisa Boone, 'They Don't Have Homes. They Don't Have Kids. Why Millennials Are Plant Addicts', www.latimes.com, 4 July 2018.
8 Dani Giannopoulos, 'Why Our Obsession with Indoor Plants Is More Important than Ever', www.domain.com.au, 7 April 2020.
9 Richard Mabey, *The Cabaret of Plants: Forty Thousand Years of Plant Life and the Human Imagination* (London, 2016).
10 Jack Goody, *The Culture of Flowers* (Cambridge, 1994).
11 Dani Nadel et al., 'Earliest Floral Grave Lining from 13,700–11,700-Y-Old Natufian Burials at Raqefet Cave, Mt Carmel, Israel', *Proceedings of the National Academy of Sciences*, CX/29 (2013), pp. 11,774–8.
12 Paul Pearce Creasman and Kei Yamamoto, 'The African Incense Trade and Its Impacts in Pharaonic Egypt', *African Archaeological Review*, XXXVI/3 (2019), pp. 347–65.
13 Judy Sund, *Exotica: A Fetish for the Foreign* (London, 2019), pp. 6–10.
14 Teresa McLean, *Medieval English Gardens* (London, 1981), p. 151.
15 Quoted in Catherine Horwood, *Potted History: The Story of Plants in the Home* (London, 2007), p. 7.
16 Celia Fiennes, *Through England on a Side Saddle in the Reign of William and Mary* (London, 1888), pp. 97–8.
17 Thomas Fairchild, *The City Gardener* (London, 1722).
18 2014 Census of Horticultural Specialities, www.nas.usda.gov, accessed 3 January 2020.

19 Chloe Blommerde, 'New Zealand's Most Expensive House Plant? $6,500 Hoya Breaks Trademe Record', www.i.stuff.co.nz, 16 June 2020.

20 Christopher Brickell and Fay Sharman, *The Vanishing Garden: A Conservation Guide to Garden Plants* (London, 1986), p. 52.

21 R. Todd Longstaffe-Gowan, 'Plant Effluvia: Changing Notions of the Effects of Plant Exhalations on Human Health in the Eighteenth and Nineteenth Centuries', *Journal of Garden History*, VII/2 (1987), pp. 176–85.

22 J. R. Mollison, *The New Practical Window Gardener: Being Practical Directions for the Cultivation of Flowering and Foliage Plants in Windows and Glazed Cases, and the Arrangement of Plants and Flowers for the Embellishment of the Household* (London, 1877), p. 52.

23 United Nations, 'Revision of World Urbanization Prospects, 68% of the World Population Projected to Live in Urban Areas by 2050', www.un.org, 18 May 2018.

24 Marc T. J. Johnson and Jason Munshi-South, 'Evolution of Life in Urban Environments', *Science*, CCCLVIII/6363 (2017).

25 Alexander Mahnert, Christine Moissl-Eichinger and Gabriele Berg, 'Microbiome Interplay: Plants Alter Microbial Abundance and Diversity within the Built Environment', *Frontiers in Microbiology* (2015), p. 887.

26 Joseph Arditti and Eloy Rodriguez, '*Dieffenbachia*: Uses, Abuses and Toxic Constituents: A Review', *Journal of Ethnopharmacology*, 5 (1982), pp. 293–302.

27 Michael G. Kenny, 'A Darker Shade of Green: Medical Botany, Homeopathy, and Cultural Politics in Interwar Germany', *Social History of Medicine*, XV (2002), pp. 481–504.

28 Nicholas C. Kawa, 'Plants that Keep the Bad Vibes Away: Boundary Maintenance and Phyto-Communicability in Urban Amazonia', *Ethnos*, LXXXVI (2020), pp. 1–17.

29 Jane Desmarais, *Monsters under Glass: A Cultural History of Hothouse Flowers from 1850 to the Present* (London, 2018).

30 N. Meeker and A. Szabari, 'From the Century of the Pods to the Century of the Plants: Plant Horror, Politics and Vegetal Ontology', *Discourse*, XXXIV/1 (2012), pp. 32–58.

31 Nathaniel Bagshaw Ward, *On the Growth of Plants in Closely Glazed Cases* (London, 1852).

32 Patrick Blanc, *The Vertical Garden* (London, 2008); Takashi Amano, www.adana.co.jp/en/contents/takashiamano, accessed 12 May 2019.

33 'Devastated Woman Discovers Plant She's Been Watering for Two Years Is Fake', www.mirror.co.uk, 3 March 2020.

34 'World's Smallest Water Lily Stolen from Kew Gardens', www.theguardian.com, 13 January 2014.

1 The Gathering of the Exotic

1 Tomas Anisko, *Plant Exploration for Longwood Gardens* (Portland, OR, 2006), p. 143.

2 IUCN Red List, www.iucnredlist.org.

3 Email communication with Bill Rotolante, January 2020.
4 Hugh Platt, *Floraes Paradise* (London, 1608).
5 Gordon Rowley, *A History of Succulent Plants* (Mill Valley, CA, 1997), pp. 43–6.
6 Rebecca Earle, 'The Day Bananas Made Their British Debut', www. theconversation.com, 10 April 2018.
7 John Gerard, *The Herball; or, Generall Historie of Plantes* (London, 1597).
8 Quoted in Douglas Chamber, 'John Evelyn and the Invention of the Heated Greenhouse', *Garden History*, XX/2 (1992), p. 201.
9 Mike Maunder, 'The Tropical Aroids: The Discovery, Introduction and Cultivation of Exotic Icons', in *Philodendron: From Pan-Latin Exotic to American Modern*, ed. Christian Larsen (Miami Beach, FL, 2015), pp. 17–31.
10 Joseph Holtum et al., 'Crassulacean Acid Metabolism in the ZZ Plant, *Zamioculcas zamiifolia* (Araceae)', *American Journal of Botany*, XCIV/10 (2007), pp. 1670–76.
11 Santiago Madriñán, *Nikolaus Joseph Jacquin's American Plants: Botanical Expedition to the Caribbean (1754–1759) and the Publication of the Selectarum Stirpium Americanarum Historia* (Leiden, 2013), p. 11.
12 Mike Maunder, 'Monstera Inc.', *Rakesprogress*, 7 (2018), pp. 220–22.
13 Donald R. Strong and Thomas S. Ray, 'Host Tree Location Behavior of a Tropical Vine (*Monstera gigantea*) by Skototropis', *Science*, 190 (1975), pp. 804–6.
14 J. López-Portillo et al., 'Hydraulic Architecture of *Monstera acuminata*: Evolutionary Consequences of the Hemiepiphytic Growth Form', *New Phytologist*, CXLV/2 (2000), pp. 289–99.
15 Tyler Whittle, *The Plant Hunters* (London, 1970), p. 118.
16 James Herbert Veitch, *Hortus Veitchii: A History of the Rise and Progress of the Nurseries of Messrs James Veitch and Sons* [1906] (Exeter, 2006).
17 Richard Steele, *An Essay upon Gardening* (York, 1793), p. 7.
18 Jianjun Chen and Richard J. Henny, 'ZZ: A Unique Tropical Ornamental Foliage Plant', *HortTechnology*, XIII/3 (2003), pp. 458–62.
19 G. Prigent, 'Huysmans Pornographe', *Romantisme*, CLXVII/1 (2015), pp. 60–75.
20 Joel T. Fry, 'America's First Poinsettia: The Introduction at Bartram's Garden', www.bartramsgarden.org, 14 December 2016.
21 Walter L. Lack, 'The Discovery, Naming and Typification of *Euphorbia pulcherrima* (Euphorbiaceae)', *Willdenowia*, XLI/2 (2011), pp. 301–9.
22 Judith M. Taylor et al., 'The Poinsettia: History and Transformation', *Chronica Horticulturae*, LI/3 (2011), pp. 23–8.
23 J. L. Clarke et al., '*Agrobacterium tumefaciens* – Mediated Transformation of Poinsettia, *Euphorbia pulcherrima*, with Virus-Derived Hairpin RNA Constructs Confers Resistance to Poinsettia Mosaic Virus', *Plant Cell Reports*, XXVII/6 (2008), pp. 1027–38.
24 Laura Trejo et al., 'Poinsettia's Wild Ancestor in the Mexican Dry Tropics: Historical, Genetic, and Environmental Evidence', *American Journal of Botany*, XCIX/7 (2012), pp. 1146–57.

25 M. T. Colinas et al., 'Cultivars of *Euphorbia pulcherrima* from Mexico', *XXIX International Horticultural Congress on Horticulture: Sustaining Lives, Livelihoods and Landscapes*, 1104 (2014), pp. 487–90.

26 Quoted in Michael Fraser and Liz Fraser, *The Smallest Kingdom* (London, 2011), p. 168.

27 Ibid., pp. 140–42.

28 Ibid., pp. 167–81.

29 Ibid., pp. 181–3.

30 D. R. Davies and C. L. Hedley, 'The Induction by Mutation of All-Year-Round Flowering in *Streptocarpus*', *Euphytica*, 24 (1975), pp. 269–75.

31 Tomas Hasing et al., 'Extensive Phenotypic Diversity in the Cultivated Florist's Gloxinia, *Sinningia speciosa* (Lodd.) Hiern, Is Derived from the Domestication of a Single Founder Population', *Plants, People, Planet*, 1/4 (2019), pp. 363–74.

2 Monsters and Beauties: Breeding a Better House Plant

1 George Gessert, *Green Light: Toward an Art of Evolution* (Cambridge, MA, 2012), p. 91.

2 Ibid., p. 26.

3 Ibid., p. 1.

4 Michael Leapman, *The Ingenious Mr Fairchild* (London, 2000).

5 Noel Kingsbury, *Hybrid: The History and Science of Plant Breeding* (Chicago, IL, 2009), p. 74.

6 Ibid., pp. 77–83.

7 Quoted ibid., p. 95.

8 Veronica M. Read, *Hippeastrum: The Gardener's Amaryllis* (Portland, OR, 2004), p. 16.

9 Ibid., p. 41.

10 Y. Wang et al., 'Revealing the Complex Genetic Structure of Cultivated Amaryllis (*Hippeastrum hybridum*) Using Transcriptome-Derived Microsatellite Markers', *Scientific Reports*, 8 (2018), pp. 1–12.

11 James Herbert Veitch, *Hortus Veitchii: A History of the Rise and Progress of the Nurseries of Messrs James Veitch and Sons* [1906] (Exeter, 2006), pp. 103–5.

12 Bodhisattva Kar, 'Historia Elastica: A Note on the Rubber Hunt in the North Eastern Frontier of British India', *Indian Historical Review*, XXXVI/1 (2009), pp. 131–50.

13 Richard J. Henny and Jianjun Chen, 'Cultivar Development of Ornamental Foliage Plants', *Plant Breeding Reviews*, 23 (2003), p. 277.

14 Ibid., pp. 278–9.

15 Ibid., pp. 271–2.

16 Hiroshi Ishizaka, 'Breeding of Fragrant Cyclamen by Interspecific Hybridization and Ion-Beam Irradiation', *Breeding Science*, LXVIII/1 (2018), pp. 25–34.

17 Dan Torre, *Cactus* (London, 2017), pp. 158–62.

18 Gideon F. Smith et al., 'Nomenclature of the Nothogenus Names × *Graptophytum* Gossot, × *Graptoveria* Gossot, and × *Pachyveria* Haage & Schmidt (Crassulaceae)', *Bradleya*, 36 (2018), pp. 33–41.

19 Jaime A. Teixeira da Silva et al., 'African Violet (*Saintpaulia ionantha* H. Wendl.): Classical Breeding and Progress in the Application of Biotechnological Techniques', *Folia Horticulturae*, XXIX/2 (2017), pp. 99–111.

20 Helene Anne Curry, *Evolution Made to Order: Plant Breeding and Technological Innovation in Twentieth-Century America* (Chicago, IL, 2016), p. 134.

21 Ibid., pp. 180–83.

22 R. J. Griesbach, 'Development of *Phalaenopsis* Orchids for the Mass-Market', *Trends in New Crops and New Uses* (2002), pp. 458–65.

23 Tim Wing Yam and Joseph Arditti, 'History of Orchid Propagation: A Mirror of the History of Biotechnology', *Plant Biotechnology Review*, III/1 (2009), pp. 1–56.

3 Health, Happiness and Mutualism

1 Emanuele Coccia, *The Life of Plants: A Metaphysics of Nature* (Cambridge, 2019), pp. 4–5.

2 United Nations, 'Revision of World Urbanization Prospects, 68% of the World Population Projected to Live in Urban Areas by 2050', www.un.org, 18 May 2018.

3 Laura J. Martin et al., 'Evolution of the Indoor Biome', *Trends in Ecology and Evolution*, XXX/4 (2015), pp. 223–32.

4 Ibid.

5 S. M. Gibbons, 'The Built Environment Is a Microbial Wasteland', *mSystems*, 1 (2016), e00033-16, doi: 10.1128/Msystems.00033-16; Gabriele Berg, Alexander Mahnert and Christine Moissl-Eichinger, 'Beneficial Effects of Plant-Associated Microbes on Indoor Microbiomes and Human Health?', *Frontiers in Microbiology*, 5 (2014), p. 15.

6 Berg, Mahnert and Moissl-Eichinger, 'Beneficial Effects'.

7 C. Neal Stewart et al., 'Houseplants as Home Health Monitors', *Science*, CCCLXI/6399 (2018), pp. 229–30.

8 Alexander Mahnert, Christine Moissl-Eichinger and Gabriele Berg, 'Microbiome Interplay: Plants Alter Microbial Abundance and Diversity within the Built Environment', *Frontiers in Microbiology*, 6 (2015), p. 887; Alexander Mahnert et al., 'Enriching Beneficial Microbial Diversity of Indoor Plants and Their Surrounding Built Environment with Biostimulants', *Frontiers in Microbiology*, 9 (2018), p. 2985; Rocel Amor Ortega et al., 'The Plant Is Crucial: Specific Composition and Function of the Phyllosphere Microbiome of Indoor Ornamentals', *FEMS Microbiology Ecology*, XCII/12 (2016).

9 United Nations, 'Revision of World Urbanization Prospects'.

10 World Health Organization, *Global Status Report on Noncommunicable Diseases 2014*, WHO/NMH/NVI/15.1 (2014).

11 Melissa R. Marselle et al., 'Review of the Mental Health and Well-Being Benefits of Biodiversity', in *Biodiversity and Health in the Face of Climate Change* (Cham, 2019), pp. 175–211; Tina Bringslimark, Terry Hartig and Grete G. Patil, 'The Psychological Benefits of Indoor Plants: A Critical Review of the Experimental Literature', *Journal of Environmental Psychology*, XXIX/4 (2009), pp. 422–33, doi: 10.1016/J.Jenvp.2009.05.001.

12 R. Todd Longstaffe-Gowan, 'Plant Effluvia: Changing Notions of the Effects of Plant Exhalations on Human Health in the Eighteenth and Nineteenth Centuries', *Journal of Garden History*, VII/2 (1987), pp. 176–85.

13 Quoted in Catherine Horwood, *Potted History: The Story of Plants in the Home* (London, 2007), p. 103.

14 Quoted ibid., p. 104.

15 Kate E. Lee et al., '40-Second Green Roof Views Sustain Attention: The Role of Micro-Breaks in Attention Restoration', *Journal of Environmental Psychology* (April 2015), pp. 182–9; Jo Barton, Murray Griffin and Jules Pretty, 'Exercise, Nature- and Socially Interactive-Based Initiatives Improve Mood and Self-Esteem in the Clinical Population', *Perspectives in Public Health*, CXXXII/2 (2012), pp. 89–96.

16 Magdalena van den Berg et al., 'Health Benefits of Green Spaces in the Living Environment: A Systematic Review of Epidemiological Studies', *Urban Forestry and Urban Greening*, XIV/4 (2015), pp. 806–16.

17 B. C. Wolverton, Anne Johnson and Keith Bounds, *Interior Landscape Plants for Indoor Air Pollution Abatement* (Davidsonville, MD, 1989).

18 Robinson Meyer, 'A Popular Benefit of Houseplants Is a Myth', *The Atlantic*, 9 March 2019; E. Cummings Bryan and Michael S. Waring, 'Potted Plants Do Not Improve Indoor Air Quality: A Review and Analysis of Reported VOC Removal Efficiencies', *Journal of Exposure Science and Environmental Epidemiology*, XXX/2 (2020), pp. 253–61.

19 Long Zhang, Ryan Routsong and Stuart E. Strand, 'Greatly Enhanced Removal of Volatile Organic Carcinogens by a Genetically Modified Houseplant, Pothos Ivy (*Epipremnum aureum*) Expressing the Mammalian Cytochrome P450 2e1 Gene', *Environmental Science and Technology*, LIII/1 (2018), pp. 325–31.

20 Susan McHugh, 'Houseplants as Fictional Subjects', in *Why Look at Plants? The Botanical Emergence in Contemporary Art* (Leiden, 2018), pp. 191–4.

21 George Orwell, *Keep the Aspidistra Flying* (London, 1956), p. 28.

22 Ernst Van Jaarsveld, *The Southern African Plectranthus* (Simons Town, 2006), pp. 72–3.

23 Orwell, *Keep the Aspidistra Flying*, p. 28.

24 Harriet Gross, *The Psychology of Gardening* (London, 2018), p. 42.

25 Chang Chia-Chen et al., 'Social Media, Nature, and Life Satisfaction: Global Evidence of the Biophilia Hypothesis', *Scientific Reports*, X/1 (2020), pp. 1–8.

26 Stephen R. Kellert, *Nature by Design: The Practice of Biophilic Design* (New Haven, CT, 2018).

27 Richard J. Jackson, Howard Frumkin and Andrew L. Dannenberg,
    eds, *Making Healthy Places: Designing and Building for Health, Well-Being, and
    Sustainability* (Portland, OR, 2012).
28 Tonia Gray, 'Re-Thinking Human–Plant Relations by Theorising Using
    Concepts of Biophilia and Animism in Workplaces', in *Reimagining
    Sustainability in Precarious Times* (Singapore, 2017), pp. 199–215.
29 Anna Wilson, Dave Kendal and Joslin L. Moore, 'Humans and
    Ornamental Plants: A Mutualism?', *Ecopsychology*, VIII/4 (2016),
    pp. 257–63.

4 The Crystal Legacy of Dr Ward

 1 Shirley Hibberd, *Rustic Adornments for Homes of Taste* (London, 1856),
    p. 135.
 2 John Claudius Loudon, *The Suburban Gardener and Villa Companion*
    (London, 1838), p. 104.
 3 Marianne Klemun, 'Live Plants on the Way: Ship, Island, Botanical
    Garden, Paradise and Container as Systemic Flexible Connected Spaces
    in Between', *Journal of History of Science and Technology*, V (Spring 2012),
    pp. 30–48; Yves-Marie Allain, *Voyages et Survie des Plantes au Temps de la Voile*
    (Paris, 2000).
 4 Allan Maconochie, 'On the Use of Glass Cases for Rearing Plants Similar
    to Those Recommended by N. B. Ward, Esq.', *Third Annual Report and
    Proceedings of the Botanical Society, Session 1838–9* (1840), pp. 96–7.
 5 Shirley Hibberd, *The Town Gardener* (London, 1855), p. 11.
 6 Nathaniel Bagshaw Ward, *On the Growth of Plants in Closely Glazed Cases*
    (London, 1842), p. 36.
 7 Stuart McCook, 'Squares of Tropic Summer: The Wardian Case,
    Victorian Horticulture, and the Logistics of Global Plant Transfer,
    1770–1910', in *Global Scientific Practice in an Age of Revolutions, 1750–1850*,
    ed. Patrick Manning and Daniel Rood (Pittsburgh, PA, 2016),
    pp. 199–215.
 8 Donal P. McCracken, *The Gardens of Empire* (London, 1997), p. 85.
 9 C. Mackay, 'The Arrival of the Primrose', *Friends Intelligencer*, XXII/8
    (29 April 1865), p. 123.
10 D. E. Allen, *The Victorian Fern Craze: A History of Pteridomania* (London, 1969).
11 Ward, *On the Growth of Plants*, p. 49.
12 Lindsay Wells, 'Close Encounters of the Wardian Kind: Terrariums and
    Pollution in the Victorian Parlor', *Victorian Studies*, LX/2 (Winter 2018),
    pp. 158–70.
13 Margaret Flanders Darby, 'Unnatural History: Ward's Glass Cases',
    Victorian Literature and Culture, XXXV/2 (2007), pp. 635–47.
14 J. Pascoe, *The Hummingbird Cabinet* (Ithaca, NY, 2006), p. 48.
15 Charles Kingsley, *Glaucus; or, The Wonders of the Shore* (London, 1890), p. 4.
16 William Scott, *The Florist's Manual* (Chicago, IL, 1899), p. 84.
17 Shirley Hibberd, *The Fern Garden* (London, 1869), p. 54.

18 Nona Maria Bellairs, *Hardy Ferns* (London, 1865), p. 77.
19 Ruth Kassinger, *Paradise under Glass* (New York, 2010), pp. 263–5.
20 Mara Polgovsky Ezcurra, 'The Future of Control: Luis Fernando Benedit's Labyrinth Series', http://post.moma.org, 4 September 2019.
21 'The Biosphere Project', www.biosphere2.org, accessed 2 March 2020.
22 Ruth Erickson, 'Into the Field', in *Mark Dion: Misadventures of a Twenty-First-Century Naturalist*, exh. cat., Institute of Contemporary Art, Boston, MA (New Haven, CT, 2017), p. 59.
23 Amy Frearson, 'Over 10,000 Plants Used to Create Grassland inside Australian Pavilion', www.dezeen.com, 26 May 2018.
24 Jacques Leenhardt, *Vertical Gardens: Bringing the City to Life* (London, 2007), p. 20.
25 Patrick Blanc, *The Vertical Garden* (London, 2008).

5 The House of Plants

1 Christian A. Larsen, ed., *Philodendron: From Pan-Latin Exotic to American Modern*, exh. cat., Florida International University/Wolfsonian Museum (Miami, FL, 2015), pp. 58–9.
2 Hannah Martin, 'The Story Behind the Iconic Banana-Leaf Pattern Design', www.architecturaldigest.com, 31 March 2019.
3 Larsen, *Philodendron*, pp. 33–112.
4 International Living Future Institute, www.living-future.org, accessed 1 April 2020.
5 Huijuan Deng and Chi Yung Jim, 'Spontaneous Plant Colonization and Bird Visits of Tropical Extensive Green Roof', *Urban Ecosystems*, XX/2 (2017), pp. 337–52.
6 See www.terreform.com, accessed 1 April 2020.
7 See www.bagarquitectura.com, accessed 6 September 2019.
8 See 'Hyun Seok-An', www.antenna.foundation, accessed 5 April 2020.
9 Tuan G. Nguyen, 'Can an Algae-Powered Lamp Quench Our Thirst for Energy?', www.smithsonianmag.com, 22 October 2013.
10 Alison Haynes et al., 'Roadside Moss Turfs in South East Australia Capture More Particulate Matter Along an Urban Gradient than a Common Native Tree Species', *Atmosphere*, X/4 (2019), p. 224.
11 See www.greencitysolutions.de, accessed 10 April 2020.
12 Emilie Chalcraft, 'Researchers Develop Biological Concrete for Moss-Covered Walls', www.dezeen.com, 3 January 2013.
13 Rima Sabina Aouf, 'Bricks Made from Waste and Loofah Could Promote Biodiversity in Cities', www.dezeen.com, 14 July 2019.
14 Amy Frearson, 'Tower of "Grown" Bio-Bricks by The Living Opens at MOMA PS1', www.dezeen.com, 1 July 2014.
15 William Myers, *Bio Art/Altered Realities* (London, 2015), p. 14.
16 Kelly Servick, 'How the Transgenic Petunia Carnage of 2017 Began', www.sciencemag.org, 24 May 2017.
17 Myers, *Bio Art*, pp. 1138–40.

18 Eleni Stavrinidou et al., 'Electronic Plants', *Science Advances*, I/I0 (6 November 2015), www.advances.sciencemag.org.

19 Mary K. Heinrich et al., 'Constructing Living Buildings: A Review of Relevant Technologies for a Novel Application of Biohybrid Robotics', *Journal of the Royal Society Interface*, 16 (2019), pp. 1–28.

20 Anne Trafton, 'Bionic Plants', https://news.mit.edu, 16 March 2014.

21 Anne Trafton, 'Nanosensor Can Alert a Smartphone When Plants Are Stressed', https://news.mit.edu, 15 April 2020.

22 Amy McDermott, 'Light-Seeking Mobile Houseplants Raise Big Questions about the Future of Technology', *Proceedings of the National Academy of Sciences*, CXVI/31 (2019), pp. 15,313–15.

23 See www.daisyginsberg.com, accessed 5 October 2019.

6 Wild and Endangered Relatives

1 Dana Goodyear, 'Succulent Smugglers Descend on California', www.newyorker.com, 12 February 2019.

2 Jared D. Margulies, 'Korean "Housewives" and "Hipsters" Are Not Driving a New Illicit Plant Trade: Complicating Consumer Motivations behind an Emergent Wildlife Trade in *Dudleya farinosa*', *Frontiers in Ecology and Evolution*, 8 (2020), p. 604,921.

3 Maurizio Sajeva, Francesco Carimi and Noel McGough, 'The Convention on International Trade in Endangered Species of Wild Fauna and Flora (CITES) and Its Role in Conservation of Cacti and Other Succulent Plants', *Functional Ecosystems and Communities*, 1/2 (2007), pp. 80–85.

4 Thomas Brooks et al., 'Global Biodiversity Conservation Priorities', *Science*, CCCXIII/5783 (2006), pp. 58–61.

5 Antonia Eastwood et al., 'The Conservation Status of *Saintpaulia*', *Curtis's Botanical Magazine*, XV/1 (1998), pp. 49–62.

6 B. L. Burtt, 'Studies in the Gesneriaceae of the Old World XXV: Additional Notes on *Saintpaulia*', *Notes of the Royal Botanic Garden Edinburgh*, XXV/3 (1964), pp. 191–5.

7 Ian Darbyshire, *Gesneriaceae, Flora of Tropical East Africa*, vol. CCXLII (London, 2006).

8 IUCN Red List, www.iucnredlist.org.

9 Ibid.

10 Dimitar Dimitrov, David Nogues-Bravo and Nikolaj Scharff, 'Why Do Tropical Mountains Support Exceptionally High Biodiversity? The Eastern Arc Mountains and the Drivers of *Saintpaulia* Diversity', *Plos One*, VII/11 (2012).

11 Dawn Edwards, 'The Conversion of *Saintpaulia*', *The Plantsman*, XVII/4 (December 2018), pp. 260–61.

12 Mark Hughes, 'The Begonia of the Socotra Archipelago', *Begonian*, 68 (November December 2001), pp. 109–213, www.begonias.org.

13 James Herbert Veitch, *Hortus Veitchii: A History of the Rise and Progress of the Nurseries of Messrs James Veitch and Sons* [1906] (Exeter, 2006).

14 James Wong, 'Gardens: All Hail the Vulcan Palm', www.guardian.co.uk, 10 January 2016.

15 Anon., 'Plant Focus: Resurrected from the Brink of Extinction', *The Plantsman* (2005/P4), p. 67.

16 Seana K. Walsh et al., 'Pollination Biology Reveals Challenges to Restoring Populations of *Brighamia insignis* (Campanulaceae), a Critically Endangered Plant Species from Hawai'i', *Flora*, 259 (October 2019).

17 Lex Thomson and Luca Braglia, 'Review of Fiji *Hibiscus* (Malvaceae-Malvoideae) Species in Section Lilibiscus', *Pacific Science*, LXXIII/1 (2019), pp. 79–121.

18 Earley Vernon Wilcox and Valentine S. Holt, *Ornamental 'Hibiscus' in Hawaii*, Bulletin no. 29, Hawaii Agricultural Experiment Station (Honolulu, HI, 1913).

19 Thomson and Braglia, 'Review'.

20 Barbara Goettsch et al., 'High Proportion of Cactus Species Threatened with Extinction', *Nature Plants*, 1/10 (2015).

21 Ana Novoa et al., 'Level of Environmental Threat Posed by Horticultural Trade in Cactaceae', *Conservation Biology*, XXXI/5 (2017), pp. 1066–75.

22 W. A. Maurice et al., '*Echinocactus grusonii*, a New Location for the Golden Barrel', *Cactusworld*, XXIV/4 (2006), pp. 169–73.

23 Rafael Ortega Varela, Zirahuen Ortega Varela and Charles Glass, 'Rescue Operations of Threatened Species in the Hydroelectric Project: Zimapán, Mexico', *British Cactus and Succulent Journal*, XV/3 (1997), pp. 123–8.

24 IUCN Red List, www.iucnredlist.org, accessed 5 March 2020.

25 Mike Maunder et al., 'Conservation of the Toromiro Tree: Case Study in the Management of a Plant Extinct in the Wild', *Conservation Biology*, XIV/5 (2000), pp. 1341–50.

26 Michael Fraser and Liz Fraser, *The Smallest Kingdom* (London, 2011).

27 Ibid.; Anthony Hitchcock, '*Erica verticillata*', www.plantzafrica.com, accessed 8 January 2019.

28 *Ladies' Floral Cabinet*, quoted in Tovah Martin, *Once upon a Windowsill: A History of Indoor Plants* (Portland, OR, 1988), p. 209.

29 Bruce Beveridge, *The Ship Magnificent*, vol. II: *Interior Design and Fitting* (London, 2009).

30 G. P. Darnell-Smith, 'The Kentia Palm Seed Industry, Lord Howe Island', in *Bulletin of Miscellaneous Information (Royal Botanic Gardens, Kew)* (London, 1929), pp. 1–4; Alba Herraiz et al., 'Developing a New Variety of Kentia Palms (*Howea forsteriana*): Up-Regulation of Cytochrome B561 and Chalcone Synthase Is Associated with Red Colouration of the Stems', *Botany Letters*, CLXV/2 (2018), pp. 241–7.

Conclusion: New Worlds

1 World Population Growth, www.ourworldindata.org, accessed 18 December 2020.

2 See 'CO2 since 1800', www.sealevel.info, accessed 18 December 2020.

3 Christopher J. Preston, *The Synthetic Age: Outdesigning Evolution, Resurrecting Species and Reengineering Our World* (Cambridge, MA, 2018).

4 Long Zhang, Ryan Routsong and Stuart E. Strand, 'Greatly Enhanced Removal of Volatile Organic Carcinogens by a Genetically Modified Houseplant, Pothos Ivy (*Epipremnum aureum*) Expressing the Mammalian Cytochrome P450 2e1 Gene', *Environmental Science and Technology*, LIII/1 (2018), pp. 325–31.

5 See www.iucn-uk-peatlandprogramme.org, accessed 2 September 2021.

6 Ferdinand Ludwig et al., 'Living Bridges Using Aerial Roots of *Ficus elastica*: An Interdisciplinary Perspective', *Scientific Reports*, IX/1 (2019), pp. 1–11; John Goddard, 'Food Preferences of Two Black Rhinoceros Populations', *African Journal of Ecology*, VI/1 (1968), pp. 1–18; Ian Kiepieland and Steven D. Johnson, 'Shift from Bird to Butterfly Pollination in *Clivia* (Amaryllidaceae)', *American Journal of Botany*, CI/1 (2014), pp. 190–200; Nicholas C. Kawa, 'Plants that Keep the Bad Vibes Away: Boundary Maintenance and Phyto-Communicability in Urban Amazonia', *Ethnos* (2020), pp. 1–17.

Bibliography

Allan, Mea, *Tom's Weeds: The Story of the Rochfords and Their House Plants* (London, 1970)

Blanc, Patrick, *The Vertical Garden* (London, 2008)

Curry, Helene Anne, *Evolution Made to Order: Plant Breeding and Technological Innovation in Twentieth-Century America* (Chicago, IL, 2016)

Desmarais, Jane, *Monsters under Glass: A Cultural History of Hothouse Flowers from 1850 to the Present* (London, 2018)

Erickson, Ruth, *Mark Dion: Misadventures of a Twenty-First-Century Naturalist*, exh. cat., Institute of Contemporary Art, Boston, MA (New Haven, CT, 2017)

Fraser, Michael, and Liz Fraser, *The Smallest Kingdom* (London, 2011)

Gessert, George, *Green Light: Toward an Art of Evolution* (Cambridge, MA, 2012)

Gross, Harriet, *The Psychology of Gardening* (London, 2018)

Horwood, Catherine, *Potted History: The Story of Plants in the Home* (London, 2007)

Jones, Margaret E., *House Plants* (London, 1962)

Kassinger, Ruth, *Paradise under Glass* (New York, 2010)

Kellert, Stephen R., *Nature by Design: The Practice of Biophilic Design* (New Haven, CT, 2018)

Kingsbury, Noel, *Hybrid: The History and Science of Plant Breeding* (Chicago, IL, 2009)

Koopowitz, Harold, *Clivias* (Seattle, WA, 2002)

Larsen, Christian A., ed., *Philodendron: From Pan-Latin Exotic to American Modern*, exh. cat., Florida International University/Wolfsonian Museum (Miami, FL, 2015)

Leapman, Michael, *The Ingenious Mr Fairchild* (London, 2000)

Leenhardt, Jacques, *Vertical Gardens: Bringing the City to Life* (London, 2007)

Mabey, Richard, *The Cabaret of Plants: Forty Thousand Years of Plant Life and the Human Imagination* (London, 2016)

Martin, Tovah, *Once upon a Windowsill: A History of Indoor Plants* (Portland, OR, 1998)

Myers, William, *Bio Art/Altered Realities* (London, 2015)

Read, Veronica M., *Hippeastrum: The Gardener's Amaryllis* (Portland, OR, 2004)

Rowley, Gordon, *A History of Succulent Plants* (Mill Valley, CA, 1997)

Staples, George W., and Derral R. Herbst, *A Tropical Garden Flora*
(Honolulu, HI, 2005)
Sund, Judy, *Exotica: A Fetish for the Foreign* (London, 2019)
Van Jaarsveld, Ernst, *The Southern African* Plectranthus (Simons Town, 2006)
Whittle, Tyler, *The Plant Hunters* (London, 1970)
Wilson, Edward O., *The Diversity of Life* (Cambridge, MA, 1992)

Associations and Websites

AFRICAN VIOLET SOCIETY OF AMERICA
www.avsa.org

AMERICAN BEGONIA SOCIETY
www.begonias.org

AMERICAN HORTICULTURAL SOCIETY
www.ahsgardening.org

BRITISH CACTUS AND SUCCULENT SOCIETY
www.society.bcss.org.uk

BRITISH STREPTOCARPUS SOCIETY
www.facebook.com/streptocarpussociety

CACTUS AND SUCCULENT SOCIETY OF AMERICA
www.cactusandsucculentsociety.org

CLIVIA SOCIETY
www.cliviasociety.org

THE GESNERIAD SOCIETY
www.gesneriadsociety.org

INTERNATIONAL AROID SOCIETY
www.aroid.org

INTERNATIONAL HIBISCUS SOCIETY
www.internationalhibiscussociety.org

NATIONAL BEGONIA SOCIETY (UK)
www.national-begonia-society.co.uk

ROYAL HORTICULTURAL SOCIETY
www.rhs.org.uk

Tropical Plant Conservation Groups

BOTANIC GARDENS CONSERVATION INTERNATIONAL
www.bgci.org

FAUNA AND FLORA INTERNATIONAL
www.fauna-flora.org

NATIONAL TROPICAL BOTANICAL GARDEN
www.ntbg.org

RAINFOREST TRUST
www.rainforesttrust.org

ROYAL BOTANIC GARDENS, KEW
www.kew.org

WORLD LAND TRUST
www.worldlandtrust.org

Acknowledgements

I have been lucky to walk between two worlds. I have had the opportunity to work in wild habitats with expert field botanists, most notably in Hawaii with my NTBG colleagues, including Ken Wood and Steve Perlman, and in eastern Africa with Quentin Luke and his colleagues from the East African Plant Red List Authority.

Similarly, I have been fortunate enough to explore the cultivated collections of commercial nurseries, botanic gardens and private collectors in the tropics and under glass. My very happy years in southern Florida and Hawaii were greatly enriched by the extraordinary network of tropical botanists and horticulturists who were generous with their time, knowledge and cuttings.

The spark for this book was ignited with two conversations: the first with Christian Larsen, then of the Wolfsonian-FIU Museum on Miami Beach; the second with Dr Rajindra Puri and his ethnobotany students at the University of Kent. Many people have helped generously with this project. Jon Drury and Matt Biggs have encouraged me throughout the process. Susyn Andrews with characteristic vigour tried to keep my nomenclature in order. Many friends and colleagues provided guidance, most notably Harvey Bernstein, Jim Folsom, Mitchell Joachim, Maureen McCadden, Craig Morell, Sara Oldfield, Brian Schrire, Paul B. Redman, Bill Rotolante, Paul Smith, Matt Taylor, Lex Thomson, John Trager and Vikash Tatayah. In Thailand I must thank Dr Weerachai Nanakorn for a fantastic tour of the Bangkok plant markets and fruit stalls. All mistakes are mine alone.

I am very grateful to the following artists who have so generously shared images of their work: Eugenio Ampudio, Keita Augstkalne, Patric Blanc, Gohar Dashti, Disney Davis and Nitin Barha, ecoLogic Studios, Laura Hart, Jamie North, Heidi Norton, Jean Nouvel, Kate Polsby, Dian Scherer, Vo Trong Nghia Architects and WOHA Studio.

Several books have provided continued inspiration. Many thanks to Catherine Horwood for *Potted History*, Tovah Martin for *Once Upon a Windowsill*, and George Staples and Derral R. Herbst for their encyclopaedic reference *A Tropical Garden Flora*. I found myself repeatedly returning to *The Smallest Kingdom* by Mike and Liz Fraser, a beautiful book.

Tribute is given to the horticulturists and conservationists working to stop plant extinctions; this is a multigenerational task as exemplified by successive teams who have nurtured *Erica verticillata* and brought it back to the wild in South Africa.

Above all, thanks to my wonderful family: Sawsan, Catherine (thanks for the editing and picture research) and Peter (thanks for the house-plant lamps); and particularly to my parents, Peter and Isabella, who taught me to love plants.

Photo Acknowledgements

❧

The author and publishers wish to express their thanks to the below sources of illustrative material and/or permission to reproduce it. Some locations of artworks are also given below, in the interest of brevity:

From the *African Violet Magazine*, III/1 (September 1949): p. 80; photo Adriano Alecchi/Mondadori via Getty Images: p. 22; courtesy of Eugenio Ampudio (photo Pedro Martínez de Albornoz): pp. 88–9; from Anon., *Ferns and Ferneries* (London, 1880): p. 113; courtesy of Ateliers Jean Nouvel (photo © Yiorgis Yerolymbos): p. 130; courtesy of Keita Augstkalne: p. 99; from Parker T. Barnes, *House Plants and How to Grow Them* (New York, 1909): p. 34; courtesy of the Barnes Foundation, Merion and Philadelphia, PA: p. 97; courtesy of Henk Beentje: p. 147; courtesy of Harvey Bernstein: pp. 40, 41, 43, 48, 159; courtesy of Patrick Blanc: p. 121; photo © The Bloomsbury Workshop, London/Bridgeman Images: p. 15; Carol M. Highsmith's America, Library of Congress, Prints and Photographs Division, Washington, DC: pp. 118–19; © CartoonStock, www.CartoonStock.com: p. 32; photo Murray Close/Sygma via Getty Images: p. 31; courtesy of CW Stockwell (photo Matt Sartain): p. 126; © Gohar Dashti, courtesy of the artist: p. 129; courtesy of ecoLogicStudio (photo © Marco Cappelletti): p. 136; © The Estate of John Nash, all rights reserved 2022/Bridgeman Images: p. 13; from *Gartenflora*, vol. I (Erlangen, 1852): p. 62; from Shirley Hibberd, *The Fern Garden: How to Make, Keep, and Enjoy it; or, Fern Culture Made Easy*, 9th edn (London, 1881): p. 109; courtesy of IKEA: p. 106 (*bottom*); photo Imagno/Getty Images: p. 169; iStock.com: pp. 11 (Kihwan Kim), 52 (leekris), 75 (EzumeImages), 83 (Praiwun), 91 (KatarzynaBialasiewicz), 106 (*top*; hamikus), 127 (FollowTheFlow), 137 (wavemovies), 149 (Nahhan); Keystone Pictures USA/ZUMA Press/Alamy Stock Photo: p. 30; photo Thomas Ledl (CC BY-SA 3.0 AT): p. 125; courtesy of Longwood Gardens, Kennett Square, PA: pp. 14, 36; © The Lucian Freud Archive/Bridgeman Images: p. 100; LuEsther T. Mertz Library, New York Botanical Garden: p. 110; courtesy of Quentin Luke: p. 67; photos Mike Maunder: p. 33; National Agricultural Library, U.S. Department of Agriculture, Beltsville, MD: pp. 53, 162; National Gallery of Art, Washington, DC: pp. 29, 57; Nationalmuseum, Stockholm (photo Cecilia Heisser): p. 10; photo © Jamie North, courtesy of the artist and Informality

Gallery: p. 122; © Heidi Norton, courtesy of the artist: p. 26; Österreichische Galerie Belvedere, Vienna: p. 107; Österreichische Nationalbibliothek, Vienna (Cod. 2773, fol. 18r): p. 6; photos Pixabay: pp. 93 (zoosnow), 157 (*top*; stux); private collection: p. 94; © Diana Scherer, courtesy of the artist: pp. 142, 143; photo Frank Scherschel/The LIFE Picture Collection/Shutterstock: p. 81; photo © 2018 Ester Segarra, reproduced with permission of Laura Hart: p. 84; Shutterstock.com: pp. 16 (*bottom*; Shuang Li), 18 (Shulevskyy Volodymyr), 19 (Svetlana Foote), 20–21 (stocksolutions), 27 (lennystan), 45 (wjarek), 49 (mayu85), 51 (Izz Hazel), 55 (Jessica Pichardo), 58 (SariMe), 59 (Anne Kitzman), 60 (mady70), 66 (Kristi Blokhin), 72 (kyrien), 78 (*top*; joloei), 78 (*bottom*; weter777), 86 (Aunyaluck), 146 (Elena-Grishina), 154 (TrishZ), 157 (*bottom*; Rob Huntley), 164 (Naaman Abreu), 168 (mathiasmoeller), 171 (Abhijeet Khedgikar); from Edward Step, *Favourite Flowers of Garden and Greenhouse*, vol. III (London and New York, 1897): p. 170; courtesy of Vikash Tatayah: p. 155; © Tate/Tate Images: pp. 128, 139; courtesy of Lex A. J. Thomson: p. 156; photo Touring Club Italiano/Marka/Universal Images Group via Getty Images: p. 98; photos Unsplash: pp. 12 (Severin Candrian), 134 (Daniel Seßler); from James H. Veitch, *Hortus Veitchii: A History of the Rise and Progress of the Nurseries of Messrs. James Veitch and Sons* (London, 1906): p. 64; courtesy of Vo Trong Nghia Architects: pp. 102, 103; photo Seana Walsh, National Tropical Botanical Garden (NTBG), Kalaheo, HI: p. 152; photo Hedda Walther/ullstein bild via Getty Images: p. 28; from N. B. Ward, *On the Growth of Plants in Closely Glazed Cases* (London, 1852): p. 108; from Robert Warner and Benjamin Samuel Williams, *The Orchid Album*, vol. I (London, 1882): p. 9; courtesy of The White Room, www.thewhiteroom.in: p. 16 (*top*); courtesy of WOHA, Singapore: pp. 132–3; courtesy of Ken Wood: p. 150.

Index

Page numbers in *italics* indicate illustrations